Green Energy and Technology

More information about this series at http://www.springer.com/series/8059

Nigel Horan · Abu Zahrim Yaser
Newati Wid
Editors

Anaerobic Digestion Processes

Applications and Effluent Treatment

Editors
Nigel Horan
Halifax
UK

Abu Zahrim Yaser
Faculty of Engineering
Universiti Malaysia Sabah
Kota Kinabalu
Malaysia

Newati Wid
Faculty of Science and Natural Resources
Universiti Malaysia Sabah
Kota Kinabalu
Malaysia

ISSN 1865-3529 ISSN 1865-3537 (electronic)
Green Energy and Technology
ISBN 978-981-13-4070-3 ISBN 978-981-10-8129-3 (eBook)
https://doi.org/10.1007/978-981-10-8129-3

Printed on acid-free paper

This Springer imprint is published by the registered company Springer Nature Singapore Pte Ltd.
part of Springer Nature
The registered company address is: 152 Beach Road, #21-01/04 Gateway East, Singapore 189721, Singapore

I wish to thank Linda, Kate and Matthew for their many years of tolerating both holiday detours to visit sewage works and guests whose conversations invariably ended up discussing sewage.

—Nigel Horan

To my mom (Hamidah Khalil), dad (Yaser Malik) and families.

Special thanks to my wife (Naimmaton Sa'adiah Mustain) and my children (A'ishah, Anas and Asma)

for endless love, support and patience.

—Abu Zahrim Yaser

This book is dedicated with much love to my husband, Razak, my parents Mama-Baba and The Wids.

—Newati Wid

Acknowledgements

The Editors gratefully acknowledge the following individuals for their time and efforts in assisting the Editors with the reviewing of manuscript. This book would not have been possible without the commitment of the Reviewers.

- Fadzil Noor Gonawan
- M. Rafatullah Lari
- Marcus Japony
- Mariani Rajin
- Md. Shafiquzzaman Siddiquee
- Md. Mizanur Rahman
- Mohd. Azmier Ahmad
- Muhd. Nazrul Hisham Zainal Alam
- Norazwina Zainol
- Nurmin Bolong
- Rafidah Hamdan
- Rahmath Abdulla
- Rosa Anna Nastro
- Rovina Kobun
- Ruwaida Abd. Rasid
- S. M. Anisuzzaman
- Wu Ta Yeong
- Zainul Akmar Zakaria

Contents

Introduction 1
N. J. Horan

Microbial Fuel Cell (MFC) Development from Anaerobic Digestion System 9
Muaz Mohd Zaini Makhtar, Mashitah Mat Don and Husnul Azan Tajarudin

Sediment Microbial Fuel Cells in Relation to Anaerobic Digestion Technology 33
Syed Zaghum Abbas and Mohd Rafatullah

A Current Review on the Application of Enzymes in Anaerobic Digestion 55
Mariani Rajin

Process Simulation of Anaerobic Digestion Process for Municipal Solid Waste Treatment 71
Noorlisa Harun, Wan Hanisah W. Ibrahim, Muhamad Faez Lukman, Muhammad Hafizuddin Mat Yusoff, Nur Fathin Shamirah Daud and Norazwina Zainol

Anaerobic Digestion of Screenings for Biogas Recovery 85
N. Wid and N. J. Horan

Anaerobic Digestion of Food Waste 105
Md. Mizanur Rahman, Yeoh Shir Lee, Fadzlita Mohd Tamiri and Melvin Gan Jet Hong

Overview of Biologically Digested Leachate Treatment Using Adsorption 123
I. Azreen and A. Y. Zahrim

Current Progress on Removal of Recalcitrance Coloured Particles from Anaerobically Treated Effluent Using Coagulation–Flocculation 149
A. Y. Zahrim

Effect of Seaweed Physical Condition for Biogas Production in an Anaerobic Digester 165
N. Bolong, H. A. Asri, N. M. Ismail and I. Saad

Phosphorus Recovery from Anaerobically Digested Liquor of Screenings 177
N. Wid

Editors and Contributors

About the Editors

Nigel Horan was a Lecturer in Public Health Engineering at the University of Leeds for 35 years during which time he published five academic text-books and with over 120 refereed technical papers. In 1998, he established Aqua Enviro Ltd. and was the company Chair until its acquisition in 2015 by Suez Advanced Water Solutions. He specialises in the treatment of wastewaters, particularly using the activated sludge process, and in the subsequent recovery of resources from sludge through anaerobic digestion.

Abu Zahrim Yaser is a Deputy Dean (Research and Innovation) at Faculty of Engineering, Universiti Malaysia Sabah (UMS). He obtained his Ph.D. from Swansea University and has published over 72 refereed technical papers. He is an outstanding reviewer for several Elsevier journals and a guest editor for *Environmental Science and Pollution Research* (Springer). He is a Visiting Scientist at the University of Hull. He is also a member of Board of Engineers (Malaysia), Institutions of Chemical Engineers (UK) and Malaysian Water Association.

Newati Wid received her Ph.D. from the University of Leeds (UK) and currently is a Senior Lecturer in Industrial Chemistry, at the Universiti Malaysia Sabah (UMS). Her research focuses on wastewater treatment and resource recovery from wastes through anaerobic digestion with over 10 refereed technical papers were published. In 2015 and 2016, she won gold medal for her research work on resource recovery at the Research and Invention Competition UMS.

Contributors

Syed Zaghum Abbas Division of Environmental Technology, School of Industrial Technology, Universiti Sains Malaysia, Penang, Malaysia

H. A. Asri Faculty of Engineering, Universiti Malaysia Sabah (UMS), Kota Kinabalu, Sabah, Malaysia

I. Azreen Chemical Engineering Programme, Faculty of Engineering, Universiti Malaysia Sabah, Kota Kinabalu, Sabah, Malaysia

N. Bolong Faculty of Engineering, Universiti Malaysia Sabah (UMS), Kota Kinabalu, Sabah, Malaysia

Nur Fathin Shamirah Daud Faculty of Chemical and Natural Resources Engineering, Universiti Malaysia Pahang, Gambang, Kuantan, Pahang, Malaysia

Mashitah Mat Don School of Chemical Engineering, Engineering Campus, Universiti Sains Malaysia, Nibong Tebal, Malaysia; Cluster of Solid Waste Management, Engineering Campus, Universiti Sains Malaysia, Nibong Tebal, Malaysia

Noorlisa Harun Faculty of Chemical and Natural Resources Engineering, Universiti Malaysia Pahang, Gambang, Kuantan, Pahang, Malaysia

Melvin Gan Jet Hong Energy Research Unit (ERU), Mechanical Engineering Programme, Faculty of Engineering, Universiti Malaysia Sabah (UMS), Sabah, Malaysia

N. J. Horan School of Civil Engineering, University of Leeds, Leeds, UK; Aqua Enviro Ltd., Wakefield, UK

Wan Hanisah W. Ibrahim Faculty of Chemical and Natural Resources Engineering, Universiti Malaysia Pahang, Gambang, Kuantan, Pahang, Malaysia

N. M. Ismail Faculty of Engineering, Universiti Malaysia Sabah (UMS), Kota Kinabalu, Sabah, Malaysia

Yeoh Shir Lee Energy Research Unit (ERU), Mechanical Engineering Programme, Faculty of Engineering, Universiti Malaysia Sabah (UMS), Sabah, Malaysia

Muhamad Faez Lukman Faculty of Chemical and Natural Resources Engineering, Universiti Malaysia Pahang, Gambang, Kuantan, Pahang, Malaysia

Muaz Mohd Zaini Makhtar School of Chemical Engineering, Engineering Campus, Universiti Sains Malaysia, Nibong Tebal, Malaysia

Mohd Rafatullah Division of Environmental Technology, School of Industrial Technology, Universiti Sains Malaysia, Penang, Malaysia

Md. Mizanur Rahman Energy Research Unit (ERU), Mechanical Engineering Programme, Faculty of Engineering, Universiti Malaysia Sabah (UMS), Kota Kinabalu, Sabah, Malaysia

Mariani Rajin Chemical Engineering Programme, Faculty of Engineering, Universiti Malaysia Sabah, Kota Kinabalu, Sabah, Malaysia

I. Saad Faculty of Engineering, Universiti Malaysia Sabah (UMS), Kota Kinabalu, Sabah, Malaysia

Husnul Azan Tajarudin Division of Bioprocess, School of Industrial Technology, Universiti Sains Malaysia, Nibong Tebal, Malaysia; Cluster of Solid Waste Management, Engineering Campus, Universiti Sains Malaysia, Nibong Tebal, Malaysia

Fadzlita Mohd Tamiri Energy Research Unit (ERU), Mechanical Engineering Programme, Faculty of Engineering, Universiti Malaysia Sabah (UMS), Sabah, Malaysia

N. Wid Faculty of Science and Natural Resources, Universiti Malaysia Sabah, Kota Kinabalu, Sabah, Malaysia

Muhammad Hafizuddin Mat Yusoff Faculty of Chemical and Natural Resources Engineering, Universiti Malaysia Pahang, Gambang, Kuantan, Pahang, Malaysia

A. Y. Zahrim Chemical Engineering Programme, Faculty of Engineering, Universiti Malaysia Sabah, Kota Kinabalu, Sabah, Malaysia

Norazwina Zainol Faculty of Chemical and Natural Resources Engineering, Universiti Malaysia Pahang, Gambang, Kuantan, Pahang, Malaysia

Introduction

N. J. Horan

Abstract Anaerobic microorganisms have a large number of biochemical pathways that allow them to oxidise organic substrates anaerobically and which can be exploited by engineers to provide valuable end-products from organic wastes. This Introduction summarises the many examples presented in this book, which are being used in Malaysia to provide sustainable solutions to organic waste generation and treatment.

Keywords Anaerobic microrganisms · Organic wastes · Malaysia
Anaerobic digestion

1 Background

Anaerobic microorganisms have evolved a large number of biochemical pathways that enable them to oxidise organic substrates in the absence of oxygen, to provide a source of carbon and energy thus ensuring their survival and reproduction. Biochemical engineers have long exploited these pathways to generate useful end products through fermentation and increasingly these skills are being applied to tackle the problems of organic wastes. When applied thoughtfully and appropriately, with due regards to process economics, anaerobic systems can reduce the mass of an organic waste and minimise its environmental impact. As a consequence, the anaerobic digestion process is well understood and sophisticated reactor design can be achieved. The process bottlenecks are recognised and solutions to free them are continually being sought. At the same time, the potential of a wide range of wastes to act as feedstocks is continually under evaluation. Methane is the most common commercially valuable end product and with good reason. However,

N. J. Horan (✉)
School of Civil Engineering, University of Leeds, Leeds, UK
e-mail: horannigel@gmail.com

N. J. Horan
Aqua Enviro Ltd., Wakefield, UK

N. Horan et al. (eds.), *Anaerobic Digestion Processes*,
Green Energy and Technology, https://doi.org/10.1007/978-981-10-8129-3_1

other high value, low volume products are receiving attention in particular highly reduced carbon end products. Such routes offer the potential of moving away from an oil-based carbon economy, delivering a wholly renewable chain from waste to product; something often referred to as the circular or green economy. Technological advances driven by this include moves away from the traditional one or two-stage stirred-tank reactors and which exploit the electrochemical properties of certain bacterial genera. These microbial fuel cells (MFCs) undergo the recognised biochemical redox pathways but with the electrons produced during the oxidative stage captured directly and exploited in a similar way to an electrical current. They have the dual advantage of waste treatment and electrical energy generation without the requirement for intermediate energy conversion technologies (Chaps. "Microbial Fuel Cell (MFC) Development from Anaerobic Digestion System" and "Sediment Microbial Fuel Cells in Relation to Anaerobic Digestion Technology"). This collection of research papers highlights the important research work currently being undertaken across the university sector in Malaysia. With a focus on improving the traditional digestion processes, for instance, by enhancing the hydrolysis stage enzymatically, they also look at the potential of the technology for tackling specific Malaysian issues such as palm oil and its associated wastes. It also covers other wastes such as wastewater screenings that are likely to increase in volume as more of Malaysia becomes sewered with the consequent uptake of downstream wastewater processing. It discusses the alternatives to conventional reactors by means of MFCs and considers the research necessary to render this approach a more mainstream technological option. This Introduction aims to place the contents of this book firmly within our existing knowledge of the anaerobic digestion process and highlight the requirements for turning these exciting research ideas into full-scale and financially viable projects.

2 Feedstocks

Anaerobic microorganisms can obtain carbon and energy from a wide range of organic compounds. The majority of feedstocks arise from naturally occurring sources and thus comprise primarily of protein, carbohydrate and fat, or lipid, all of which are readily digestible. At the same time, a range of synthetic organic materials of industrial origin are also amenable to digestion. Whereas the former is predominantly particulate with a dry solid content of 6% upwards, the latter is largely soluble and thus measured in terms of its chemical oxygen demand (COD). Anaerobic digestion as a preliminary stage can often be a cost-effective option for rendering such industrial wastes benign, particularly if the COD is in excess of 5000 mg/l.

Perhaps the most widely researched feedstock for anaerobic digestion is sewage sludge, both primary (the solids that settle out from the wastewater in primary settlement tanks) and secondary (the excess microorganisms produced during the biological treatment stage and removed during a second settlement process).

Primary sludge is predominantly faecal organic matter and paper fibres and, since it has already received preliminary digestion in the human gut, the organic matter digests relatively quickly. By contrast, secondary sludge comprises largely of bacteria and protozoa produced on a trickling filter (humus sludge) or from the activated sludge process (waste activated sludge) together with a sticky exopolysaccharide matrix produced by these organisms. It digests more slowly as the microbial cell walls resist hydrolysis. The importance of the hydrolysis stage as potentially rate-limiting has long been recognised. Consequently, it is common practice now to accelerate the digestion process and enhance the volume of methane produced through the use of a preliminary hydrolysis stage. This is usually thermal, operating at temperatures and pressures of 165 °C and 6 bar or biological achieved by operating at a pH value of 5.0–5.5 which provides optimal conditions for the hydrolytic bacteria. There are other biological alternatives to these options, and more recent approaches to enhancing the hydrolysis stage using enzymes are discussed in Chap. "A Current Review on the Application of Enzymes in Anaerobic Digestion". In view of the long history of anaerobic digestion and our clear understanding of it, it is unsurprising that a number of mathematical models of the process are available and that they are finding increasing application in design and operation. A review of this approach is provided in Chap. "Process Simulation of Anaerobic Digestion Process for Municipal Solid Waste Treatment", which considers in detail the application of one particular model.

An essential feature of a municipal biological wastewater treatment process is that the catchment has a sewerage network to convey wastewater to the treatment plant. These sewerage networks can be susceptible to blockage with fats oils and greases (FOG) particularly where a large number of catering establishments are found in a concentrated area. The increasing use of disposable wipes that are disposed down the toilet enhances the formation of FOG and huge agglomerations, known as fatbergs, can accumulate. A fatberg of over 140 tonnes was recently found in the London's sewers, which took a team of eight people 21 days to clear using high-pressure hoses. This material is very amenable to anaerobic digestion with high methane yields. Preventing its accumulation in the sewers can be achieved through a widespread use of grease traps and the coordinated collection of grease trap fat for digestion offers an economic solution to minimise fatberg formation. Another by-product of the sewage treatment process is the screenings collected from the inlet screens during preliminary treatment. This material is largely organic and is traditionally disposed of to landfill. As Chap. "Anaerobic Digestion of Screenings for Biogas Recovery" describes, this is another waste that is very amenable to digestion with a high methane yield and with a large carbon footprint saving through diverting from landfill to the anaerobic digestion route.

The skills gained from digestion of sewage sludge have been applied to other feedstocks and digestion of farm wastes, food wastes and municipal organic wastes is now common practice in many countries. Food wastes are particularly amenable to anaerobic digestion with up to 90% of the volatile solids being converted to methane gas. As food wastes generally have a volatile solids concentration of >70%, then a large reduction in the mass of solids can be achieved. It is conventional practice to incorporate a pasteurisation stage to ensure complete pathogen destruction and temperatures of 70 °C held for 30 min in plug flow reactors, will ensure that the digestate is free from human, animal and plant pathogens. When compared to other disposal routes such as landfill, the environmental benefits are large. However, it is generally accepted that in order to operate a successful digester an annual feedstock of at least 30,000 tonnes is required and so in this guise, the technology is of little use to smaller communities. Chapter "Anaerobic Digestion of Food Waste" outlines the research being undertaken to scale down the process and thus provides digestion at the household level upwards.

Although digesters are now being used to divert and treat food waste destined for landfill, there is a historic legacy of landfilling. These landfills over time will produce a leachate as they saturate from precipitation. This leachate can have a high organic content often in excess of 20,000 mg/l COD and is highly coloured. Adsorption is one option to remove the colour and it can be applied either pre- or post the digestion process (Chap. "Overview on Biologically Digested Leachate Treatment Using Adsorption").

A big challenge that has so far resisted research efforts is in the digestion of cellulosic crop residues. The complex structure of the cellulose molecule protects it from enzymic hydrolysis and it requires complex pretreatment using, for instance, heat or high pH. Even after such treatment, methane yields are too low to make the process of economic interest. Although the global quantities of crop residues are huge, they are spread over a vast area and generally only for a restricted growing season each year. Few undertakings generate adequate volumes to warrant construction of storage, pretreatment and digestion facilities. They do however offer the advantage of a ready land bank for digestate recycling. Consequently, as well as the technological challenge, there is an equally important economic one that requires any solution be appropriate to the scale of the wastes available. Malaysia is unique globally in the amount of palm oil it produces and the refining of this oil at the palm mill generates a palm oil mill effluent (POME). Treating strategies for treating this POME and in particular for removing its colour are discussed in Chap. "Current Progress on Removal of Recalcitrance Coloured Particles from Anaerobically Treated Effluent Using Coagulation-Flocculation".

Although specifically a crop, seaweed is a naturally occurring alga that is found in large and often nuisance causing quantities. It is known to digest anaerobically with a reasonable methane yield and its harvesting from the ocean can form part of a phosphorus treatment and recovery scheme. Chapter "Effect of Seaweed Physical Condition for Biogas Production in an Anaerobic Digester" provides the latest thinking on how this material can be best exploited and the potential for resource recovery from it.

All these processes share common problems and often there are transferrable solutions, although dialogue between the respective practitioners is often lacking or sparse. Typical problems encountered are that (i) although anaerobic digestion will destroy much of the organic matter (as much as 90% destruction for food wastes and around 50% for sewage sludge), there is a very little reduction in volume.

This is because digesters are typically fed at a dry solids concentration of 6–14% and so the majority of the feed is water. Indeed for some wastes that arrive at the digester with solids concentrations >14%, dilution with water may be required, and thus, there is a greater volume of waste leaving the system. Economic and secure routes for final recycling of this material (known as biosolids or digestate) must be available; (ii) the digestion process is poor at pathogen destruction (this includes human, animal and plant pathogens), and this will impact on the potential routes for final recycling. A thermal stage such as pasteurisation will ensure complete elimination of these pathogens. If well thought out and designed, it can also enhance hydrolysis and product yield, reducing the operating costs and making a contribution to capital costs.

3 End Products

The take-up of anaerobic digestion is often encouraged for a number of different reasons and in particular for its ability to reduce the mass of biodegradable organic mass and divert this material from landfill, whilst at the same time reducing the carbon footprint of the waste. But for the operator, it is often the potential value of the end products that is the main driver for technology take-up. Manipulation of the digester operating conditions permits the production of a wide range of potentially valuable, highly reduced carbon end products, but methane is by far the most popular of these. Its biggest advantage is the ease with which it can be separated from the liquid fraction in the digester without recourse to sophisticated separation techniques. It can also be used to generated heat and power at a very small scale, and combined heat and power units driven by methane fuel are becoming much smaller. Indeed, they are now of a size small enough for an individual dwelling where they can provide all the necessary heat and power.

All organic wastes will contain phosphorus at varying concentrations and much of the particulate phosphorus is released during digestion and can be recovered in a soluble form in the sludge dewatering liquor that results from thickening and dewatering the digester contents prior to land recycling. At present at domestic treatment plants, this dewatering liquor is recirculated back to the head of the treatment plant or may be used to dilute the feedstock at food waste digesters. In both cases, this raises the risk of struvite precipitation in the pipework. Options for phosphorus removal and recovery form the subject for Chap. “Phosphorus Recovery from Anaerobically Digested Liquor of Screenings”.

There is a large global research effort into generation of more novel products. These include butanol and acetate both of which are precursors to biodegradable plastic synthesis. Separation of the product from the digester contents is a major challenge and the capital costs of the distillation technology require that large quantities of feedstock are available in order to make the investment worthwhile. At present, product yields upwards of 500,000 tonnes of waste per annum are required for a profitable process and this must be available within a reasonable travel distance from the facility. Certain crop wastes may provide the necessary quantities within reasonable distances and as mentioned earlier, seaweed may prove a perfect feed for this option (Chap. "Effect of Seaweed Physical Condition for Biogas Production in an Anaerobic Digester").

4 Environmental Benefits

Often the main driver for the take-up of anaerobic technologies is the environmental benefits it can offer a country or region and its importance in helping meet national environmental targets. The role of legislation and subsidies are crucially important in encouraging companies and industry to invest in the technology. Legislation is also important in driving technological advances and legislators have a key role in helping mould the technology to the specific needs of their own country. Historically, the water companies operated almost all the digesters in the UK and they were used to stabilise sewage sludge, before recycling it to agricultural land. The sole role of the digestion process was to reduce the volatile solids content of the sludge, thus reducing odour during recycling. Consequently, these digesters were generally operated with long retention times and without heating. The low cost of energy and the absence of appropriate combined heat and power technology (CHP), meant that there was no interest in collecting and using the methane generated. However, rising energy costs and development of suitable CHP engines meant that methane could be collected and used economically at the larger sites. This potential was increased further by generous government subsidies for renewable energy, and this innovation saw a widespread increase in the number of digesters built or upgraded to produce methane. Public fears about the potential health risks due to human and animal pathogens from food grown on land treated with digested sewage sludge, led to legislation requiring a pasteurisation stage in the digestion process, pretreating the feedstock to a temperature of at least 70 °C for 30 min. This in turn led to the uptake of thermal hydrolysis and it is now standard practice at larger works to employ a thermal hydrolysis stage prior to mesophilic digestion.

The food waste sector saw a similar rise in the uptake of digestors, this time in response to legislation to divert organic wastes away from landfill. Although composting was an alternative route for biodegradable organic waste, it carried no subsidies for the process and the only sources of income were a gate fee for the waste received and the sale of the composted product. By contrast, digesters offered the multiple benefits of producing renewable energy and reducing the amount of CO_2 emissions as well as producing a digestate that could be recycled to agricultural land. As a result, there are now more digesters treating food waste than sewage sludge, a trend that took place in less than a decade.

Clearly, the importance of government and regulators in driving technological innovation should not be underestimated, and an awareness of potential relevant future legislation can offer major commercial advantages.

5 Economic Models

Successful take-up of anaerobic digestion across a region or country, requires a robust financial model to ensure that this process is the lowest cost environmental option. Ideally, the income from the digester end products should cover the operating costs and also pay off the capital costs over the lifetime of the facility. However, this is rarely possible without subsidies of some kind. As an example, a food waste digester receives a gate fee for every 1 m^3 of food waste that it receives. At the same time, it can sell the energy it produces in the form of methane and receives income based on the electricity and heat that this produces. It may also receive a fee for any digestate generated. However, this income alone is not enough to operate the facility profitably and this relies on energy subsidies for the renewable electricity and heat it produces. These subsidies are usually higher than the actual income from the sale of electricity and heat and are guaranteed for the life of the facility (although the actual financial value of the subsidy may vary in response to market demands). Subsidies can take many forms. As well as subsidies for the generation of renewable heat and energy, they might also involve payments for savings in CO_2 emissions, reduction of organic waste going to landfill and recovery of nutrients such as phosphorus. In those areas where the soil is very low in organic matter, subsidies based on the amount of digestate produced and recycled to agricultural land should prove very effective.

6 Conclusions

Anaerobic digestion has been practiced for over a century and now provides an opportunity to recycle organic wastes and reduce greenhouse gas emissions by capturing the carbon in the waste either as methane or as other highly reduced and commercially end products. At the same time, the nutrients present in the waste can be recycled to agricultural land either as digestate or by direct phosphorus capture. This book places digestion in a Malaysian context. It looks at potential waste feedstocks such as POME, seaweed and screenings that are readily found in Malaysia and presents a digester model that can be used to predict likely performance. It examines future treatment alternatives such as MFCs, adsorption and enzymic technologies. It is clear from this book that despite its long history, digestion still has a long way to go and through the application of dedicated researchers will find an important role to play in achieving environmental improvements, delivering renewable energy and reducing GHG emissions in Malaysia.

Microbial Fuel Cell (MFC) Development from Anaerobic Digestion System

Muaz Mohd Zaini Makhtar, Mashitah Mat Don and Husnul Azan Tajarudin

Abstract Alternative energy technologies become more attractive as the price of energy from fossil fuels becomes more expensive and the environmental concerns from their use amount. The microbial fuel cell (MFC) is an innovative renewable energy technology that also serves to treat wastewater through the bacteria-driven oxidation of organic substrates. The electrons from oxidizing organic substrates are shuttled from the anode to the cathode, producing a current. The only byproducts of this process are respiratory waste in the form of water and carbon dioxide. MFCs operated using mixed cultures currently achieve substantially greater power densities than those with pure cultures. Community analysis of the microorganism that exists in MFCs has so far revealed a great diversity in composition. MFCs are being constructed using a variety of materials and diversity of configurations. These systems are operated under a range of conditions that include differences in pH, electrode distance, moisture content and temperature. The MFCs are believed to be a promising technology which could be implemented as wastewater treatment, recover energy production method from biomass, environment sensor and product recovery method. There are several aspects need to be considered in order to have an efficient upscale MFC.

Keywords Fuel · Microbial fuel cell · Electron

M. M. Z. Makhtar · M. M. Don
School of Chemical Engineering, Engineering Campus, Universiti Sains Malaysia, 14300 Nibong Tebal, Malaysia

H. A. Tajarudin (✉)
Division of Bioprocess, School of Industrial Technology, Universiti Sains Malaysia, 11800 Nibong Tebal, Malaysia
e-mail: azan@usm.my

M. M. Don · H. A. Tajarudin
Cluster of Solid Waste Management, Engineering Campus, Universiti Sains Malaysia, 14000 Nibong Tebal, Malaysia

N. Horan et al. (eds.), *Anaerobic Digestion Processes*,
Green Energy and Technology, https://doi.org/10.1007/978-981-10-8129-3_2

1 Introduction

Energy policy is currently a high-profile issue in many countries, with a focus on energy security, sources of fossil fuels and alternative clean energy [1]. Oil and gas prices have increased significantly and remain uncertain. The main sources of energy used around the world at this time derive from fossil fuels, but there are many issues associated with their use [2]. Fossil fuels are not a sustainable source of energy and will one day be completely exhausted [3]. Energy-saving technologies need to be developed to conserve oil reserves and at the same time, more sustainable technologies are needed in the next decade [4, 5]. A sustainable energy specialization involving carbon technology and renewable energy should be developed. Reliance on renewable energy is growing with technologies such as solar, wind and biomass energy playing an increasingly important role in meeting our energy future [6].

Microbial fuel cells (MFCs), which convert biochemical to electrical energy, can be part of this. MFCs can be used in energy production based on biomass through the anaerobic digestion system [3, 7, 8]. Regarding this, microbial fuel cells (MFC) are bioelectrical devices that harness the natural metabolism of exoelectrogenic bacteria to produce electrical energy [9]. They have recently become more attractive because the substrate used in such systems can be almost any biodegradable organic matter, including domestic [10] and industrial wastewater [3]. Only exoelectrogenic bacteria are capable of transferring metabolically generated electrons across their cell membranes to an external electrode [11]. By using anaerobic digestion in MFCs, there is a great potential for both renewable energy generation and waste remediation, as follows:

(a) It is clean energy and reduces the offset cost in terms of electricity (no aeration and recirculation processes).
(b) Reduces the cost of sludge treatment (anaerobic digestion reduces the formation of sludge). Aerobic process produces a larger quantity of waste sludge.
(c) Investment cost is lower than aerobic treatment since two technologies are not required, that is, (i) system to handle the water and (ii) an additional system to separate and handle the biosolid.
(d) Anaerobic digestion offers the potential for good nutrient.

2 Microbial Fuel Cell System

The first step in MFCs is that the anaerobic exoelectrogenic bacteria in the anodic chamber (anoxic condition) begin to oxidize the added substrate (S) and release electrons (e^-) towards the anode, as well as the protons (H^+). Carbon dioxide (CO_2) is produced as a product of oxidation. The electrons produced are transported from the anode (A) to the cathode (C) through the external circuit to generate electricity.

After passing the proton exchange membrane (PEM), the protons enter the cathode where oxygen (O_2) reduction occurs and water (H_2O) is formed. Figure 1 shows the schematic diagram of typical MFC [12, 13]. Further explanations are provided in Sects. 2.1 and 2.2.

The reaction at the anode and cathode electrode in a typical MFC using acetate (CH_3COO^-) as a model substrate model is presented below [14]:

Anodic reaction: Acetate oxidation

$$CH_3COO^- + 2H_2O \rightarrow 2CO_2 + 7H^+ + 8e^- \quad (1)$$

Cathodic reaction: Oxygen reduction

$$O_2 + 4e^- + 4H^+ \rightarrow 2H_2O \quad (2)$$

The MFC acts as a galvanic cell in which the anodic potential (E_{an}) is lower than the potential of the cathode (E_{cat}). Cell reactions occur spontaneously and as a result, electrical current is generated.

2.1 *Biological Concept in MFC*

Bacteria obtain their carbon sources from a variety of organic compounds. Carbon and energy can be obtained from processing organics such as lipids, proteins and carbohydrates [15]. In complex redox reactions, organic substrates act as an electron donor resulting in the production of the energy carrier molecules (ATP). Figure 2 illustrates the glycolysis and citric acid cycle processes where electron carriers (NADH and $FADH_2$) were formed. Lipids, carbohydrates and protein undergo different reaction sequences to convert them via glycolysis and related processes to the acetyl unit of acetyl CoA. This molecule will pass through the citric acid cycle, where oxidation reactions occur to decrease NAD^+ and FAD to form the electron carriers, NADH and $FADH_2$ (Fig. 2). The citric acid cycle takes place in

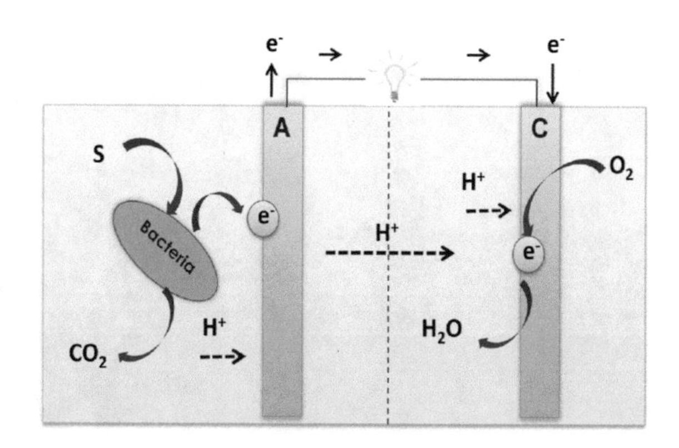

Fig. 1 Schematic diagram of a typical MFC [14]

the cytoplasm, and then, the electron carriers NADH and $FADH_2$ transfer the electron to the membrane layer [16]. In fact, before it is sent to the terminal electron acceptor (oxygen or any of reduced inorganic compounds), the electron needs to go through different membrane proteins. Some of them pump protons once they are reduced. This provides energy through a proton gradient and mediated through the ATP synthase transmembrane protein. The energy used by the cell to produce ATP (chemical energy for living organism) results from the phosphorylation process of ADP. Figure 3 presents a schematic representation of bacterial membrane respiration, and the number of components of the electron transport chain varies with species. The reduction of the terminal electron receptor generates the ATP and this process is called respiration [17]. In MFCs, bacteria can replace the electron as terminal electron receptors at the anodic compartment in the reactor.

2.2 *Chemical Concept in MFC*

The difference in redox potential between two distinct electrodes is the key to determining the MFC voltage [19]. Microbes grow by forming the colonies and start to respire on the surface of the anode. At the same time, highly reduced biomolecules begin to accumulate and surround the anode. The build-up of metabolic byproducts results in a decrease of electrical potential (typically around 0.1 to −0.4 V compared to a standard hydrogen electrode). The cathode is placed in an

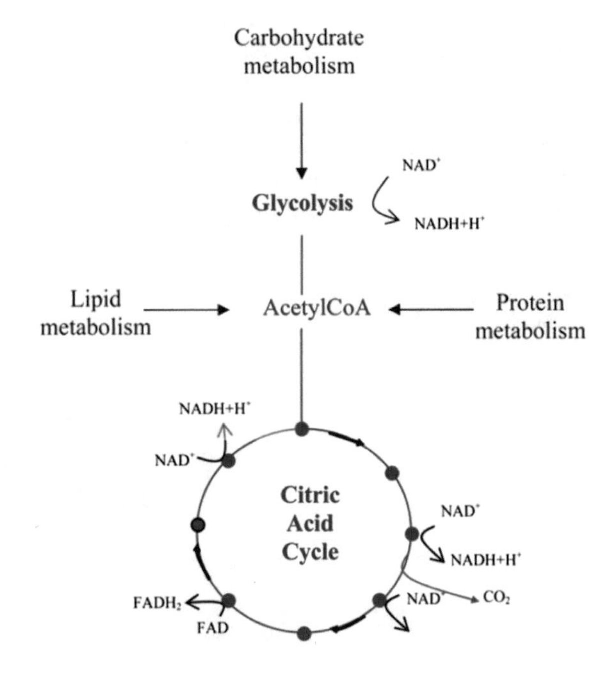

Fig. 2 Glycolysis and citric acid cycle processes where electron carriers (NADH and $FADH_2$) form [18]

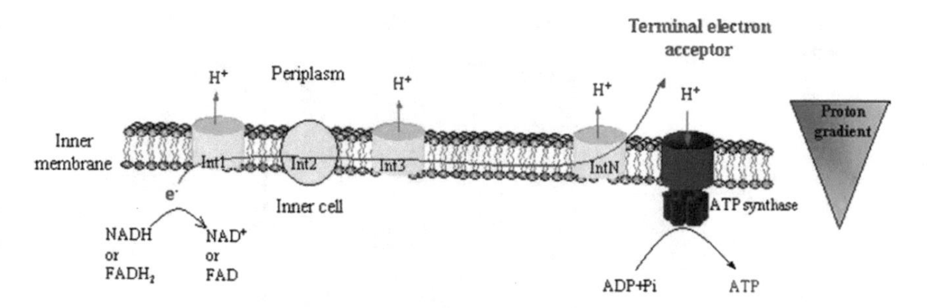

Fig. 3 Schematic representation of bacterial membrane respiration, and the number of components of the electron transport chain varies with species [18]

enriched oxygen environment and this situation creates a higher electrical potential (commonly around 0.4–0.8 V compared to an SHE) [20]. Thus, the working voltage can be calculated by subtracting the anode potential from the cathode potential. The maximum voltage that can be achieved between two electrodes is approximately 1.2 V (theoretically) because the minimum potential of reduced molecules is −0.4 V versus SHE and the redox potential of oxygen is 0.8 V. Figure 4 shows the respiratory chain and standard potential. That is how the voltage being determined depends on the location of the exit chain of respiratory enzyme. Table 1 represents MFC electrode reactions and corresponding redox potential. There are many places in the electron transport chain (ETC) that electrons can exit from the system [19].

2.3 *Characteristic for Anode Respiring Bacteria in the MFC*

Investigation in the field of bioelectricity production has focused on the identification and isolation of bacteria that have the ability to transfer electrons to the electrode [22]. Some studies have shown that metabolic activity is most similar to electron transfer in MFCs, specifically dissimilatory metal reduction. Electrons from the anaerobic exoelectrogenic bacteria accepted by the anode, act as a terminal electron acceptor in the anoxic environment (absence of oxygen) [23]. The bacteria shunt away electrons produced from the oxidation of organic compounds. *Shewanella* is the first organism reported to transfer electrons to the electrode surface [24]. There were three methods to transfer the electron to the anode; (1) direct transfer from the cell walls of microbes to the anode surface, (2) using secondary biomolecules to shuttle electrons to the anode or (3) the transfer of electrons through conductive appendages, called 'nanowires', planted by microbes [16, 19, 23]. Figure 5 shows the electron transport in microbial fuel cells. Recent research has found evidence that electrons could also be transferred from one microbe to interspecies using the conductive appendages [23]. Clearly, more information is needed on interspecies electron transfer.

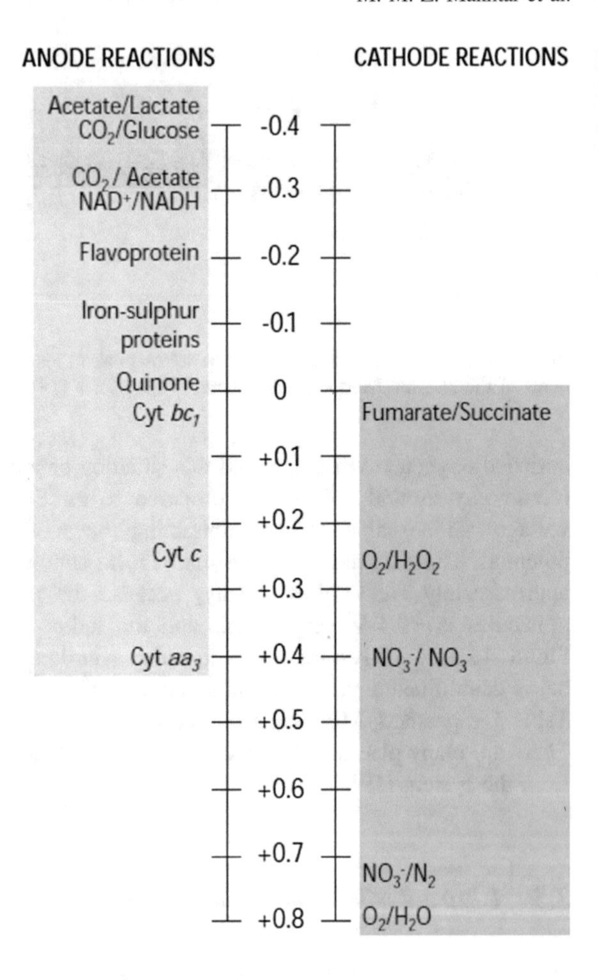

Fig. 4 The respiratory chain and standard potential [21]

2.3.1 Axenic Bacterial Culture

MFC can be operated using axenic bacterial culture or in mixed culture. Axenic bacterial describes MFCs operating with a culture in which only a single species is present. Some species of bacteria in MFCs, where the metal-reducing bacteria are the most important, have the ability to directly transfer electrons to the anode (Table 2). The single species of EB in MFCs can typically be easily be found in sediments where they utilize Fe (III) and Mn (IV) as the insoluble electron acceptors. *Shewanella putrefaciens* has a special feature where there is specific cytochrome outside the cell membrane that allows electrochemical activity in case it is grown under anaerobic conditions. The family of *Geobacteraceae* also have the same capability to survive in an aerobic environment and show that they formed a layer of biofilm at the anode in the MFC to transfer electrons from acetate with high efficiency [26].

Table 1 MFC electrode reactions and corresponding redox potential [21]

Oxidation/reduction pairs	E^o (mV)
H^+/H^2	−420
$NAD^+/NADH$	−320
S^0/HS^-	−270
SO_4^{2-}/H_2S	−220
$Pyruvate^{2-}/Lactate^{2-}$	−185
2,6-AQDS/2,6-AHQDS	−184
$FAD/FADH_2$	−180
Menaquinone ox/red	−75
Pyocyanin	−34
Humic substances oc/red	−200 to +300
Methylene blue ox/red	+11
$Fumarate^{2-}/Succinate$	+31
Thionine ox/red	+64
Cytochrome b (Fe^{3+})/Cytochrome b (Fe^{2+})	+75
Fe(III)EDTA/Fe(II) EDTA	+96
Ubiquinone ox/red	+113
Cytochrome c (Fe^{3+})/Cytochrome c (Fe^{2+})	+254
O_2/H_2O_2	+275
Fe(III) citrate/Fe(II) citrate	+372
Fe(III) NTA/Fe(II) NTA	+385
NO_3^-/NO^2	+421
$Fe(CN)_6^{3-}/Fe(CN)_6$	+430
NO_2^-/NH_4^+	+440
O_2/H_2O	+820

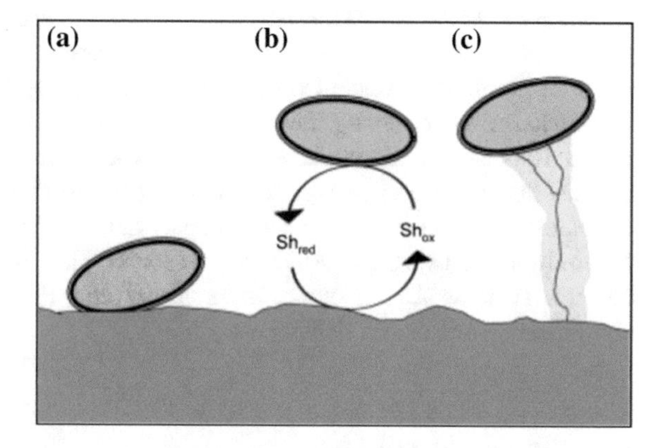

Fig. 5 Electron transport in microbial fuel cells: **a** direct electron transfer, **b** an electron shuttling and **c** solid conductive appendages, called 'nanowires' [25]

In anoxic sediment, *Rhodoferax* species have a high efficiency in transferring electrons to the graphite anode using glucose as the carbon source [29]. This was the first bacterial strain reported capable of completely mineralizing glucose to CO_2 and at the same time generating electricity at 90% efficiency [14, 30]. From the

Table 2 Overview of metal-reducing bacteria applied in MFCs

Organism	References
Shewanella putrefaciens	[14]
Geobacter sulfurreducens	[27]
Geobacter metallireducens	[28]
Desulfuromonas acetoxidans	[28]
Rhodoferax ferrireducens	[29]

perspective of performance, bacterial species such *S. putrefaciens*, *Geobacter sulfurreducens* and *Rhodoferax ferrireducens* were able to produce current densities in the range of 0.2–0.6 mA and 1–17 mW power density of surface/m^2 in conventional (woven) graphite electrodes [29]. These bacteria generally showed a high efficiency of electron transfer but the disadvantages of axenic culture are that they had slow growth rate and high substrate specificity (especially for acetate or lactate) [25]. In addition, compared to mixed culture, they are relatively low efficiency in terms of energy transfer. Furthermore, the likelihood of contamination with unwanted bacteria inside the MFCs is also one of the major obstacles when dealing with a pure culture [26].

2.3.2 Mixed Bacterial Culture

MFCs using mixed bacterial cultures have a number of significant advantages over MFCs using pure culture. They have much higher resistance to interference process, higher substrate intake rates, low substrate specificity and higher power generation [16, 31]. Electrochemically active mixed culture mostly comes from the sediment of seas and lakes or from the activated sludge at wastewater treatment plants. Species of electrochemically active bacteria include *Geobacteraceae*, *Desulfuromonas*, *Alcaligenes faecalis*, *Enterococcus faecium*, *Pseudomonas aeruginosa*, *Proteobacteria*, *Clostridia*, *Bacteroides* and *Aeromonas* species. The study of Kim's research group also concluded that the nitrogen-fixing bacteria (e.g. *Azoarcus* and *Azospirillum*) were among the electrochemically active bacterial populations [30]. While Rabaey and co-workers in their experimental work showed an electrochemically active consortium can be obtained from sludge that went through a methanogenesis process, after continuously harvested the anodic populations over a 5-month period using glucose as carbon source [25]. It was a facultative anaerobic bacteria such as *Alcaligenes*, *Enterococcus* and *Pseudomonas* species. Table 3 shows the power output delivered by MFC.

2.4 Electron Acceptor

The main aim of the MFC is to generate power; thus, the electron transport mechanisms are investigated to increase this power. A good electron acceptor must

Table 3 Overview of the power generated by the MFC using axenic and mixed culture [32]

Microorganism	Substrate	Anode	Current (mA)	Power (mW/m^2)
Shewanella putrefaciens	Lactate	Woven graphite	0.031	0.19
Geobacter sulfurreducens	Acetate	Graphite	0.41	13
Rhodoferax ferrireducens	Glucose	Woven graphite	0.2	8
Mixed seawater culture	Glucose	Porous graphite	74	33
	Acetate	Graphite	0.23	10
	Sulfide/acetate	Graphite	60	32
Mixed active sludge culture	Acetate	Graphite	5	–
	Glucose	Graphite	30	3600
	Sewage	Woven graphite	0.2	8

have a high redox potential, exhibit fast kinetics, be materially cost-effective, and if possible use sustainable materials and tools where available [33, 34]. Currently, various compounds have been used as the MFC cathode electron receiver. Among these compounds, two categories were used solely for power generation which are better with higher catholytes energy and oxygen [21].

2.4.1 High-Energy Catholytes

In early investigations of MFC, ferricyanide was a common choice due to its high redox potential, easy operation and fast kinetics. However, it is not suitable for wider applications and as a long-term cathode because it is unsustainable and requires higher maintenance with regular replenishment [25]. For the investigation of MFC design, ferricyanide has been frequently selected as a standard, followed by the anode bacteria, substrate, electrode materials and membrane materials. Despite this ferricyanide has become a benchmark to evaluate various electron acceptors. Another high-energy catholyte is permanganate. It has been used in the brush reactor and H-type reactor and it recorded a higher power generation compared to the ferricyanide. In addition to these, neutral red, thonin, methyl-viologen and anthraquinone-2,6, disulfonate are also capable to be the high-energy catholytes. To explore and optimize the output electricity and operating conditions, these high-energy catholytes could be used. The main obstacles in using them to achieve scalable devices are that they need to be maintained frequently as mentioned before.

2.4.2 Oxygen

Oxygen is the most promising electron acceptor due to the high redox potential of the oxygen reduction reaction (ORR), its abundance, high availability and low cost. Researchers aimed to improve the ORR by enlarging the effective surface area of the cathode, applying higher pressure to decrease concentration loss, using a membrane-cathode assembly (MCA) to increase the rate of proton transfer and using a catalyst to boost the reaction kinetics [17, 23].

3 Types of MFC

There is intense competition to find the best material and system configuration of MFCs to achieve a high performance and these the main objectives for every laboratory study. They hope to improve the voltage output, coulombic efficiency, stability and longevity. Besides aiming for the performances, the practicality of the MFC in terms of cost for the materials and reactor architecture also becomes a priority in the real application [35].

3.1 Aqueous-Cathode MFC with PEM

The basic MFC design (Fig. 1) consists of two chambers, anode and cathode, separated by a membrane called a proton exchange membrane (PEM). A carbon felt electrode, aqueous medium and bacterial cultures were placed inside the anode chamber [36]. While the cathode chamber consisted of a platinized carbon cloth electrode with a buffer solution. MFC operates two respiration processes at the same time, aerobic and anaerobic in the cathode and anode, respectively, thus generating electricity [15]. Oxygen concentration plays an important factor in the performance of MFCs. By sparging pure oxygen into the cathode, the dissolved oxygen in the cathode solution is increased and power generation improved. By contrast, at the anode chamber, oxygen was totally removed by sparging nitrogen gas [37]. PEM membranes are used mainly in two separate MFCs (H-type) as a way to block the oxygen mass transfer between the liquid-filled anode and cathode chamber [38]. The solution at the cathode containing dissolved oxygen cannot be allowed to mix with the solution of bacteria in the anode chamber. However, the H-type MFC reactor faced several problems involving proton transfer efficiency of PEM due to biofouling and oxygen leakage [35].

The study of Chae and co-workers proved that the leakage of oxygen from the cathode to the anode is because the membrane PEM, Nafion, was permeable to the oxygen. The characteristics of Nafion PEMs for oxygen mass transfer coefficient (K_o) and the diffusion coefficient of oxygen (D_o) were 2.80×10^{-4} and 5.35×10^{-6} cm^2/s, respectively, when 50 mM phosphate buffer was used as the catholyte [39]. These researchers also stated that biofilm would attach to the PEMs after operating over long periods of time, thus affecting the efficiency of mass transport through the membrane [40].

3.2 *Air-Cathode MFC Without PEM*

Air-cathode MFCs have become popular among researchers as they are economical and scalable, which can be practically implemented on a large scale such as at wastewater treatment plant, plus the MFC materials are affordable [41]. An air-cathode chamber MFC was proposed by Liu and Logan; they found the low coulombic result is due to the presence of a membrane [42] that allows oxygen to diffuse into the anode. It clearly showed that measurement of the oxygen flow rate oxygen diffusion into the anode chamber with PEM is much lower than the diffusion of oxygen into the anode with PEM removed [39]. Oxygen may be used by bacteria that grow in cathode instead of the anode in the absence of the PEM. Loss of substrate oxidation due to aerobic bacteria on the cathode instead of the anode had lowered the power generation of the membrane-less MFC [20]. The usage of an air-cathode single chamber membrane-less MFC exposing the cathode directly to air and water was a good idea because air sparging was no longer needed. Theoretically, an aerobic biofilm formed on the internal cathode surface (the surface facing the anode) and thus removed any oxygen that entered the chamber, maintaining anaerobic conditions in the anode chamber [43, 44]. However, this process does not guarantee the possibility of an aerobic anode.

4 Design of MFC

Researchers have proposed various scalable designs to build an MFC. In most studies, a configuration commonly used is the traditional one, which was dual chambered (H-shaped) MFC, where two bottles or chambers are connected by tubes containing the membrane separator [38]. Initially, the salt bridge in the reactor is used as an ion exchange channel between the anode and cathode, but this was later replaced with a membrane cation/proton exchange [45]. The use of air or aqueous cathode by the single-chamber fuel cell generated a higher power density than the conventional two-chambered model [41]. Many modifications of the existing models have recently been implemented to improve the power density and maintain consistency during the production. Some of the better known designs are represented in Fig. 6 which represents diagrams for typical microbial fuel cell (MFC): (a) Single-chamber MFC, (b) double-chamber MFC, (c) tubular-chamber MFC, (d) stack MFC and (e) flat-plat MFC. Whereas the basic components of MFC are shown in Table 4.

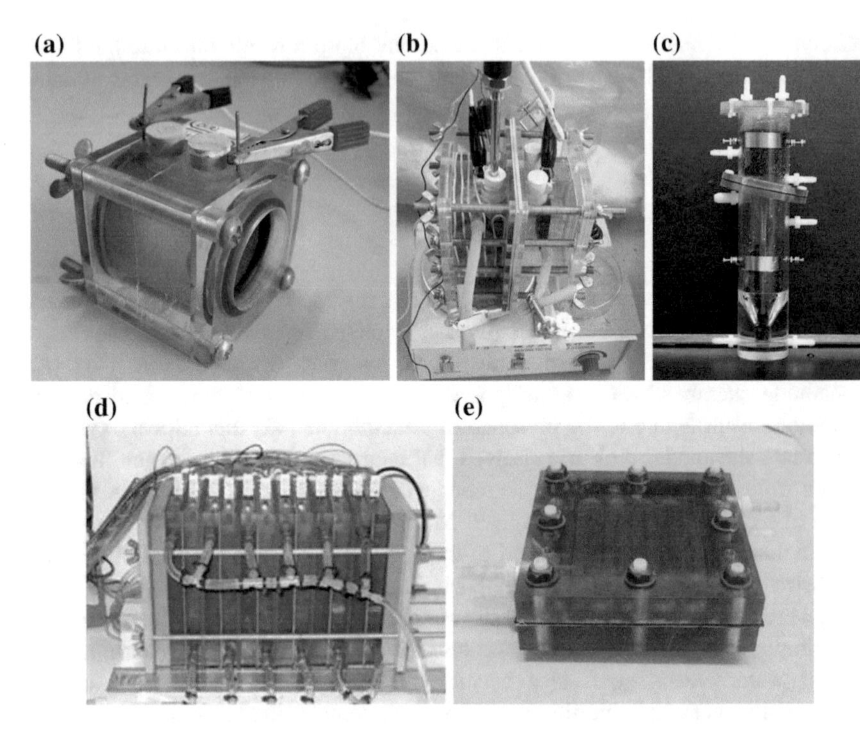

Fig. 6 Diagrams for typical microbial fuel cell (MFC): **a** single-chamber MFC, **b** double-chamber MFC, **c** tubular-chamber MFC, **d** stack MFC and **e** flat-plat MFC [41]

5 Power Density Measurement

To determine the performance of the MFC reactor in the production of cell voltage or electricity, the system needs to be optimized for power production. The power generated is related to the voltage obtained, E_{MFC}, across the external load and the current by $P = IE_{MFC}$. While current and external resistance, R_{ext}, is related by $I = E_{MFC}/R_{ext}$. Power can also be represented as [9, 11]

$$P = \frac{E_{MFC}^2}{R_{ext}} \quad (3)$$

Power density can be measured using several methods. There are power densities normalized by the surface area of the anode, cathode surface area, surface area or volume of a liquid membrane reactor. The power should be normalized in the unit of volume or area, so it easy to use the value to compare with other MFC power generation results.

Table 4 Basic components of MFC [11]

Items	Materials	Remarks
Anode	Pt, Pt black, graphite, graphite felt, carbon cloth, carbon paper, reticulated vitreous carbon (RVC)	Necessary
Cathode	Pt, Pt black, graphite, graphite felt, carbon cloth, carbon paper, RVC	Necessary
Anodic chamber	Polycarbonate, glass, Plexiglas	Necessary
Cathodic chamber	Polycarbonate, glass, Plexiglas	Optional
Proton exchange system	Nafion, polyethylene, Ultrex, salt bridge	Necessary
Electrode catalyst	MnO_2, Fe^{3+}, Pt, Pt black	Necessary

5.1 Power Output Normalized by Electrode Surface Area

This is the typical measurement method used by the researchers to calculate power [9, 11]:

$$P_A = \frac{E_{MFC}^2}{(R_{ext}A_{An})} \tag{4}$$

The power generated is related to the anode surface area available for microbial growth. But this principle is not applicable for all MFC systems architecture. For example, there was only one side of the anode being used after being pressed to the surface. Both sides of the anode surface area may be used if the anode is suspended or exposed to the bacterium. For a case where there was a very high surface area of the anode compared to the cathode, it is more suitable to normalize the power generation by the surface area of the cathode.

5.2 Power Output Normalized by Membrane Surface Area

The membrane surface area is used when the MFC has two chambers in which the anode and cathode are separated by the membrane. Oh and Logan stated that there are significant effects on the sizes of the anode, cathode and PEM to the power production. The power density represented as [9, 11]:

$$P_A = \frac{E_{MFC}^2}{(R_{ext}A_{PEM})} \tag{5}$$

5.3 Power Output Normalized by Volume

Power production was normalized using the anode volume (based solely on geometry) or the anode liquid volume or total reactor volume (both the cathode and anode chambers). For the case where the cells are cultured in a separate flask outside the reactor and are recirculated through the anode, this volume may also be included. A volumetric power density based on the anode reactor volume is [9, 11]:

$$P_A = \frac{E_{MFC}^2}{(R_{ext} V_{An})} \tag{6}$$

6 Environmental Variables Affecting Electricity Generation

Both physical and chemical variables of the incubation process conditions of MFC can influence electricity generation because exoelectrogenic bacteria are often sensitive to any changes in environmental conditions. Therefore, it becomes desirable to operate the process under optimum conditions and consequently optimize the performance of the process. Hence, the effects of environmental conditions on electricity generation MFC are discussed below:

6.1 pH

The pH is believed to have a great impact on the activity of exoelectrogenic bacteria. Generally, it has been suggested that high electricity generation was obtained when the EB worked best at the optimum pH. Therefore, in order to obtain the highest production, the incubation process needs to be operated at the optimum pH. The pH will change the internal resistance of the MFC. Internal resistance decreases with increasing pH difference between anode and cathode solution. A higher pH difference between anode and cathode affects the current as it increased the proton flux rate [46]. Lower pH could also affect the performance of MFC in producing current for electricity generation. A less acidic environment is less suitable as a base environment in MFC. This is due to acidic products of fermentation that would decrease the power production of electricity [47].

6.2 Electrode Distance

A great deal of investigations have been conducted that have demonstrated the electrode distance effect on electricity generation. Increasing the distance between the electrodes causes a slight decrease in the voltage because the longer distance leads to a greater internal resistance. These cause protons from the anode to take a longer time to transfer to the cathode and complete the circuit. This phenomenon is called ohmic loss [48]. The ohmic losses can be reduced by reducing the electrode spacing, hence inducing greater efficiency of the MFC. Although further decreasing the anode and cathode distance reduced the resistance, the power decreased because of oxygen crossover from the cathode to the anode [49]. It is suggested that MFC should be constructed by placing electrodes as close as possible to increase power output because the electron transfer through the external circuit to the cathode might be the limiting factor for the electricity production [50].

6.3 Temperature

Temperature is one of the main factors known to impact both the metabolism and the activity of EB in the MFC. Temperature is a critical variable in MFC incubation because its functions to maintain cell viability and production efficiency. In most cases, an optimal temperature is different for exoelectrogenic bacteria growth, electricity generation and COD removal. MFC studies are normally conducted at elevated temperature of 30–37 °C since high temperature can accelerate the intra-cellular biochemical reaction rate and the growth rate of the bacteria. Acceleration of metabolic rate results in rapid bacterial growth. Larrosa-Guerrero et al. (2010) studied the effect of temperature ranging from 4 to 35 °C on the performance of MFC. They found that an increase in temperature will increase electricity generation. However, temperatures that are too high do not guarantee higher voltage. The bacterial components in the cell, such as protein, nucleic acids and others are temperature-sensitive and may suffer irreversible damage, leading to cellular damage or even death [51].

7 Future Application of MFC

7.1 Wastewater Treatment

More than 2 billion people around the world do not have adequate sanitation [52] and the main cause is due to lack of funds to construct and maintain water treatment facilities [34]. Every year in the U.S., about $25 billion is spent on clean water and wastewater treatment, and it is expected that the overall cost will become approximately >$2 trillion including the cost for building, maintaining and operating the system [53]. There is a huge energy reserve in the wastewater that goes unnoticed. This is in the form of biodegradable organic matter and the aim should be to recover that energy. It is reported that a conventional wastewater treatment plant in Toronto, Canada containing about 9.3 times more energy than was used to treat the wastewater [54]. While a study by Logan stated that processing wastewater for domestic, animal and food uses approximately consist of 17 GW. This amount is equivalent to the energy needed to supply the whole water infrastructure in the U. S. It is a lot of energy and if the energy could be recovered, it would mean the treatment plant could run using its own energy supply [40].

Many substrates have been tested as the fuel in the MFC system including a variety of wastewater types [32, 55]. The application of MFC in wastewater treatment has several advantages over existing processes such as during the process for energy recovery as electricity, where less excess sludge was produced under more stable conditions compared to aerobic treatment. The conventional wastewater treatment plant spent a large amount of money on sludge disposal due to the huge quantities of excess sludge production [53]. For example, in a conventional WWTP designed to receive a daily influent flow of 5400 m^3 containing about 500 mg/L concentration of biodegradable chemical oxygen demand (bCOD), this means an influx organic matter of 2700 kg dry weight per day with a formation of sludge at 0.4 g cell dry weight per g bCOD. So every day, about 1080 kg of sludge was formed, and the cost to dispose it off could be up to €500 per tonne dry matter [25]. It is expected that much higher cost will be needed as the cost of operation, aeration and pumping are not yet calculated. The difference between aerobic processes and the MFC system is that during the aerobic process, the microorganisms use all the energy from oxidation of organic matter, whereas for the MFC system, the microorganism uses only a small amount of energy for growth, and the rest is directly converted into electricity. Figure 7 presents energy conversion in an MFC. There are several methods to recover energy, including sludge incineration, sludge gasification, pyrolysis of sludge, fermentation or anaerobic digestion [56]. The advantages and technical status are summarized in Table 5.

7.2 Renewable Energy Production from Biomass

Production of renewable energy from biomass waste may be more viable for the foreseeable future. Agricultural residues provide an abundance of renewable lignocellulosic materials and are a promising feedstock for the MFC. But the microorganisms inside the MFC cannot directly consume lignocellulosic biomass as they need to be broken down through the hydrolysis process and turned into simple sugars such as monosaccharides or other low molecular weight compounds [57]. Extraction of soluble sugar using steam explosion is the most effective treatment process. According to research conducted by Catal and his friends, they stated that all monosaccharides were good substrates for the MFC to generate electricity [58]. In other research, usage of neutral hydrolyzate from the steam explosion of corn stover as a substrate in the MFC system had yielded a total 933 mW/m^2. Therefore, MFC technology seems technically feasible to recover energy from this and other biomass waste materials. A large variety of substrates

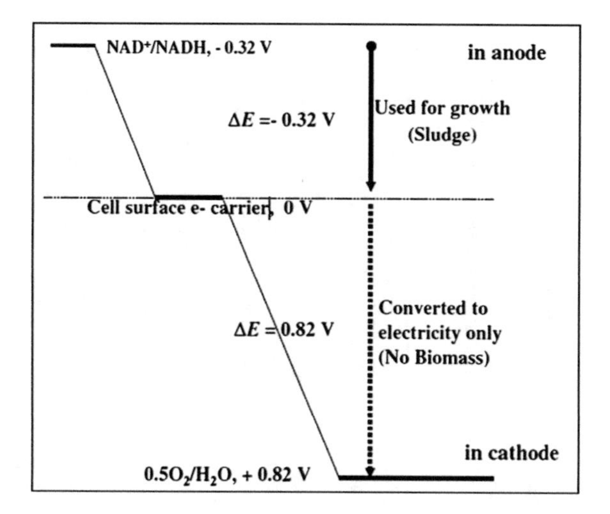

Fig. 7 Energy conversion in an MFC [25]

can be used inside the MFC as the fuel source including starch and all types of wastewater (domestic, industrial and artificial). They are presented in Table 6.

7.3 Environmental Sensor

MFCs can be implemented to power a device that collects data on the natural environment which is located in rivers and sea. The device sensor needs enough power supply for its daily operation and it is difficult to routinely access the system to replace the battery. As a consequence, researchers have developed sediment fuel cells that can be operated under these environments. The fuel of the MFC was organic matter contained in the sediment. The power generated is low at <30 mW/m^2 due to the low organic concentration and high internal resistance. However, a modification to the circuit of the sediment fuel cell, which allows the energy being stored and the low power density to be offset and able to release data in bursts to central sensors. Figure 8 shows the sediment microbial fuel cell called a benthic unattended generator (BUG) powering a data buoy that monitored humidity, pressure and the temperature of air and water.

7.4 Product Recovery

Various reactions involving organic and inorganic species can be undertaken by the modified MFC. Fischer and his co-worker reported that most sewage sludge remained toxic even though it had gone through the treatment processes. Many countries forbid its use as a fertilizer or for direct application in agriculture.

Table 5 Typical capacity and efficiency of clean energy technology to recover wastewater [50]

Conversion option	Technology	Types	Typical capacity	Net efficiency	Status
Thermal process	Sludge combustion	Heat	Domestic boiler (1–5 MW)	Modern furnace 70–90% domestic fireplace	The conventional furnace still used widely in Europe
		CHP	0.1–1 MW	69–90%	Widely applied in Scandinavia, Australia
		Stand-alone	1–10 MW 20–100 MW	80–100% 20–40% (electricity)	Conventional technology in Scandinavia-wide
		Co-combustion	5–20 MW (electricity)	30–40	Commonly used by EU countries. Recent year, mostly considered as air pollution
Biological process	Sludge gasification	Heat	5–20 MW (electricity)	10–15 heat content	Limitation: relatively high operating course, energy product
	Sludge pyrolysis	Bio-oil	Several 100 kW	Several MW electricity	Not commercialize in Australia, Germany and France
	Anaerobic digestion (bioelectricity)	Landfill gas	100 kW (electricity)	Depend on gas engine efficient	Widely applied at EU
		Microbial electrolysis cell	110 W/m^3 depend on H_2 cell fuel cell efficiency	82% (UHV basis)	Produced hydrogen through electrolysis in biological reactor but the technology is still not possible to apply
		Microbial fuel cell	55–1600 W/m^3	20–80% coulombic efficiency as the basis	**Promising technology where generating electricity while treating the wastewater**

Table 6 MFCs with different substrates and the maximum current produced [32]

Types of substrate	Concentration	Source inoculums	Type of MFC	Current density (mA/ cm^2)
Acetate	1 g/L	Pre-acclimated bacteria from MFC	Cube shape one chamber MFC with graphite fibre brush anode (7170 m^2/m^3 brush volume)	0.8
Corn stover biomass	1 g/L	Domestic wastewater	One chamber membrane-less air-cathode MFC with carbon paper anode (7.1 cm^2) and carbon cloth electrode	0.15
Landfill leachate	6000 mg/L	Leachate and sludge	Two-chambered MFC with carbon veil electrode (30 cm^2)	0.0004
Starch	10 g/L	*Pure culture of Clostridium butyricum*	Two-chambered MFC with woven graphite anode (7 cm^2) and ferricyanide catholyte	1.3
Domestic wastewater	600 mg/L	Anaerobic sludge	Two-chambered mediator-less MFC with plain graphite electrode (50 cm^2)	0.06

However, an important nutrient element for plant growth, phosphorus, is trapped inside the sludge. Then, MFC can solve the problem by using a two-chamber MFC with the *Escherichia coli* K12 as the biocatalyst, and the electron being donated via methylene blue in the anode chamber. The sewage sludge was heated and pulverized at the cathode chamber and the electron and proton from the cathode were used to liberate phosphate in the form of phosphoric acid from ferric phosphate hydrate (starting material). There was 3% of ferric phosphate by mass in the sewage sludge and the MFC was able to recover about 82% yields with approximately 30 mW/m^2 power generated within 21 days of operations. Finally, the phosphate could be obtained from treatment of the recovered phosphoric acid with magnesium chloride.

The same process was used by Heijne and co-worker during the recovery of copper. 'Microbial fuel cell component' once again played a role as a source of electrons for the recovery step and it took place at the cathode. An acidic copper chloride solution was supplied to reduce the pure copper and lastly, it deposited on the cathode surface. Within 1 week of operation, a high power density (800 mW/m^2) was reported with almost perfect recovery of copper. Once again with the same procedure, a group of researchers led by Choi managed a perfect recovery from tetrachloroaurate with 6.5 W/m^2 power density generation. The group also tested silver nitrate to extract the silver, and the result was 4 W/m^2. The achievement of

Fig. 8 Sediment microbial fuel cell called benthic unattended generator (BUG)

having two successes (recovery of precious metal and high power density) at the same time proved that MFC could be a promising metal recovery device. In other words, if the metal was present in the wastewater, the dual task could be accomplished by the MFC.

8 Conclusion

The MFCs using anaerobic digestion is a promising device for the future with functions in energy and environmental applications. However, MFC still struggles to be implemented on a large scale due to several limitations and challenges, namely, cost of materials and difficulties in scalability to plant scale. Currently, cheap alternatives for electrode and membrane are the main aim of the researchers. There are some reasonable ideas for the future such as low power generation for industrial applications, the usage of the power storage device could make it possible to boost the power output during discharge. Furthermore, in rural areas, it is also possible to fabricate homemade MFC for daily usage to power low energy devices. Finally, exploration of MFC is needed to fulfil the aim of powering the facilities at wastewater treatment plant by using the electricity gained from the waste itself, thus resulting in energy saving.

References

1. Baicha Z, Salar-García MJ, Ortiz-Martínez VM, Hernández-Fernández FJ, de los Ríos AP, Labjar N, Lotfi E, Elmahi M (2016) A critical review on microalgae as an alternative source for bioenergy production: a promising low cost substrate for microbial fuel cells. Fuel Process Technol 154:104–116
2. Sen S, Ganguly S (2017) Opportunities, barriers and issues with renewable energy development—a discussion. Renew Sustain Energy Rev 69:1170–1181
3. Gude VG (2016) Wastewater treatment in microbial fuel cells—an overview. J Clean Prod 122:287–307
4. Ahmed S, Islam MT, Karim MA, Karim NM (2014) Exploitation of renewable energy for sustainable development and overcoming power crisis in Bangladesh. Renew Energy 72:223–235
5. Branker K, Pathak M, Pearce JM (2011) A review of solar photovoltaic levelized cost of electricity. Renew Sustain Energy Rev 15:4470–4482
6. Tommasi T, Lombardelli G (2017) Energy sustainability of Microbial Fuel Cell (MFC): a case study. J Power Sources 356:438–447
7. Guo X, Zhan Y, Chen C, Cai B, Wang Y, Guo S (2016) Influence of packing material characteristics on the performance of microbial fuel cells using petroleum refinery wastewater as fuel. Renew Energy 87, Part 1:437–444
8. Logan BE (2012) Essential data and techniques for conducting microbial fuel cell and other types of bioelectrochemical system experiments. ChemSusChem 5:988–994
9. Ren L, Ahn Y, Logan BE (2014) A two-stage microbial fuel cell and anaerobic fluidized bed membrane bioreactor (MFC-AFMBR) system for effective domestic wastewater treatment. Environ Sci Technol 48:4199–4206
10. Lau LC, Tan KT, Lee KT, Mohamed AR (2009) A comparative study on the energy policies in Japan and Malaysia in fulfilling their nations' obligations towards the Kyoto Protocol. Energy Policy 37:4771–4778
11. Li W-W, Yu H-Q, He Z (2014) Towards sustainable wastewater treatment by using microbial fuel cells-centered technologies. Energy Environ Sci 7:911–924
12. Saratale GD, Saratale RG, Shahid MK, Zhen G, Kumar G, Shin H-S, Choi Y-G, Kim S-H (2017) A comprehensive overview on electro-active biofilms, role of exo-electrogens and their microbial niches in microbial fuel cells (MFCs). Chemosphere 178:534–547
13. Saba B, Christy AD, Yu Z, Co AC (2017) Sustainable power generation from bacterio-algal microbial fuel cells (MFCs): an overview. Renew Sustain Energy Rev 73:75–84
14. Lovley DR (2008) The microbe electric: conversion of organic matter to electricity. Curr Opin Biotechnol 19:564–571
15. Thomas YRJ, Picot M, Carer A, Berder O, Sentieys O, Barrière F (2013) A single sediment-microbial fuel cell powering a wireless telecommunication system. J Power Sources 241:703–708
16. Wang G, Wei L, Cao C, Su M, Shen J (2017) Novel resolution-contrast method employed for investigating electron transfer mechanism of the mixed bacteria microbial fuel cell. Int J Hydrogen Energy 42:11614–11621
17. Zhang Y, Jiang J, Zhao Q, Gao Y, Wang K, Ding J, Yu H, Yao Y (2017) Accelerating anodic biofilms formation and electron transfer in microbial fuel cells: role of anionic biosurfactants and mechanism. Bioelectrochemistry 117:48–56
18. Schaetzle O, Barriere F, Baronian K (2008) Bacteria and yeasts as catalysts in microbial fuel cells: electron transfer from micro-organisms to electrodes for green electricity. Energy Environ Sci 1:607–620
19. Zhi W, Ge Z, He Z, Zhang H (2014) Methods for understanding microbial community structures and functions in microbial fuel cells: a review. Bioresour Technol 171:461–468
20. Cheng S, Logan BE (2011) Increasing power generation for scaling up single-chamber air cathode microbial fuel cells. Bioresour Technol 102:4468–4473

21. Watanabe K (2008) Recent developments in microbial fuel cell technologies for sustainable bioenergy. J Biosci Bioeng 106:528–536
22. Zou L, Qiao Y, Zhong C, Li CM (2017) Enabling fast electron transfer through both bacterial outer-membrane redox centers and endogenous electron mediators by polyaniline hybridized large-mesoporous carbon anode for high-performance microbial fuel cells. Electrochim Acta 229:31–38
23. He C-S, Mu Z-X, Yang H-Y, Wang Y-Z, Mu Y, Yu H-Q (2015) Electron acceptors for energy generation in microbial fuel cells fed with wastewaters: a mini-review. Chemosphere 140:12–17
24. Zheng X, Chen Y, Wang X, Wu J (2017) Using mixed sludge-derived short-chain fatty acids enhances power generation of microbial fuel cells. Energy Proc 105:1282–1288
25. Rabaey K, Verstraete W (2005) Microbial fuel cells: novel biotechnology for energy generation. Trends Biotechnol 23:291–298
26. Roy S, Marzorati S, Schievano A, Pant D (2017) Microbial fuel cells A2. In: Abraham MA (ed) Encyclopedia of sustainable technologies. Elsevier, Oxford, pp 245–259
27. Nancharaiah YV, Venkata Mohan S, Lens PNL (2015) Metals removal and recovery in bioelectrochemical systems: a review. Bioresour Technol 195:102–114
28. Bond DR, Holmes DE, Tender LM, Lovley DR (2002) Electrode-reducing microorganisms that harvest energy from marine sediments. Science 295:483–485
29. Chaudhuri SK, Lovley DR (2003) Electricity generation by direct oxidation of glucose in mediatorless microbial fuel cells. Nat Biotech 21:1229–1232
30. Kim J, Min B, Logan B (2005) Evaluation of procedures to acclimate a microbial fuel cell for electricity production. Appl Microbiol Biotechnol 68:23–30
31. Santoro C, Arbizzani C, Erable B, Ieropoulos I (2017) Microbial fuel cells: from fundamentals to applications. J Power Sources: A Rev 356:225–244
32. Pant D, Van Bogaert G, Diels L, Vanbroekhoven K (2010) A review of the substrates used in microbial fuel cells (MFCs) for sustainable energy production. Bioresour Technol 101:1533–1543
33. Harnisch F, Aulenta F, Schröder U (2011) 6.49 - Microbial fuel cells and bioelectrochemical systems: industrial and environmental biotechnologies based on extracellular electron transfer. In: Editor-in-Chief: Murray M-Y (ed) Comprehensive biotechnology, 2nd edn. Academic Press, Burlington, pp 643–659
34. Du Z, Li H, Gu T (2007) A state of the art review on microbial fuel cells: a promising technology for wastewater treatment and bioenergy. Biotechnol Adv 25:464–482
35. Janicek A, Fan Y, Liu H (2014) Design of microbial fuel cells for practical application: a review and analysis of scale-up studies. Biofuels 5:79–92
36. Zhang Q, Hu J, Lee D-J (2016) Microbial fuel cells as pollutant treatment units: research updates. Bioresour Technol 217:121–128
37. Ben Liew K, Daud WRW, Ghasemi M, Leong JX, Su Lim S, Ismail M (2014) Non-Pt catalyst as oxygen reduction reaction in microbial fuel cells: a review. Int J Hydrogen Energy 39:4870–4883
38. Logan BE, Regan JM (2006) Electricity-producing bacterial communities in microbial fuel cells. Trends Microbiol 14:512–518
39. Chae KJ, Choi M, Ajayi FF, Park W, Chang IS (2008) Mass transport through a proton exchange membrane (Nafion) in microbial fuel cells. Energy Fuels 22:169–176
40. Kim JR, Cheng S, Oh S-E, Logan BE (2007) Power generation using different cation, anion, and ultrafiltration membranes in microbial fuel cells. Environ Sci Technol 41:1004–1009
41. Logan BE, Hamelers B, Rozendal R, Schröder U, Keller J, Freguia S, Aelterman P, Verstraete W, Rabaey K (2006) Microbial fuel cells: methodology and technology†. Environ Sci Technol 40:5181–5192
42. Liu H, Logan BE (2004) Electricity generation using an air-cathode single chamber microbial fuel cell in the presence and absence of a proton exchange membrane. Environ Sci Technol 38:4040–4046

43. Kiely PD, Cusick R, Call DF, Selembo PA, Regan JM, Logan BE (2011) Anode microbial communities produced by changing from microbial fuel cell to microbial electrolysis cell operation using two different wastewaters. Bioresour Technol 102:388–394
44. Jang JK, Pham TH, Chang IS, Kang KH, Moon H, Cho KS, Kim BH (2004) Construction and operation of a novel mediator- and membrane-less microbial fuel cell. Process Biochem 39:1007–1012
45. Min B, Logan BE (2004) Continuous electricity generation from domestic wastewater and organic substrates in a flat plate microbial fuel cell. Environ Sci Technol 38:5809–5814
46. Langergraber G, Muellegger E (2005) Ecological sanitation—a way to solve global sanitation problems? Environ Int 31:433–444
47. Klingel F, Montangero A, Koné D, and Strauss M (2002) Fecal sludge management in developing countries. In: A planning manual, 1st edn. Duebendorf/Accra
48. Shizas I, Bagley D (2004) Experimental determination of energy content of unknown organics in municipal wastewater streams. J Energy Eng 130:45–53
49. Choudhury P, Uday USP, Mahata N, Nath Tiwari O, Narayan Ray R, Kanti Bandyopadhyay T, Bhunia B (2017) Performance improvement of microbial fuel cells for waste water treatment along with value addition: a review on past achievements and recent perspectives. Renew Sustain Energy Rev 79:372–389
50. Oh ST, Kim JR, Premier GC, Lee TH, Kim C, Sloan WT (2010) Sustainable wastewater treatment: how might microbial fuel cells contribute. Biotechnol Adv 28:871–881
51. Sun Y, Cheng J (2002) Hydrolysis of lignocellulosic materials for ethanol production: a review. Bioresour Technol 83:1–11
52. Catal T, Li K, Bermek H, Liu H (2008) Electricity production from twelve monosaccharides using microbial fuel cells. J Power Sources 175:196–200
53. Zuo Y, Maness P-C, Logan BE (2006) Electricity production from steam-exploded corn stover biomass. Energy Fuels 20:1716–1721
54. Reimers CE, Tender LM, Fertig S, Wang W (2000) Harvesting energy from the marine sediment–water interface. Environ Sci Technol 35:192–195
55. Fischer F, Bastian C, Happe M, Mabillard E, Schmidt N (2011) Microbial fuel cell enables phosphate recovery from digested sewage sludge as struvite. Bioresour Technol 102:5824–5830
56. Heijne AT, Liu F, van der Weijden R, Weijma J, Buisman CJN, Hamelers HVM (2010) Copper recovery combined with electricity production in a microbial fuel cell. Environ Sci Technol 44:4376–4381
57. Choi C, Hu N (2013) The modeling of gold recovery from tetrachloroaurate wastewater using a microbial fuel cell. Bioresour Technol 133:589–598
58. Choi C, Cui Y (2012) Recovery of silver from wastewater coupled with power generation using a microbial fuel cell. Bioresour Technol 107:522–525

Sediment Microbial Fuel Cells in Relation to Anaerobic Digestion Technology

Syed Zaghum Abbas and Mohd Rafatullah

Abstract An anaerobic sediment microbial fuel cell (SMFC) is a device that with the help of microbial catalytic activities, simultaneously bioremediate pollutants and transfers chemical energy into electricity. SMFC attracts the attention of many researchers due to its mild operating conditions. In SMFC operation, exoelectrogens and electrotrophs are mostly involved. Although there is the great capacity of SMFC such as an alternative energy source, a biosensor for pollutants and oxygen, and a novel wastewater treatment system, high optimization is needed to accomplish the maximum microbial potential. Power output and Coulombic efficiency are significantly affected by the diversity of microbes in the anodic chamber of an SMFC, design of the SMFC and operational conditions. Until now, real-world applications of SMFC have been limited because of their low power density level of several thousand mW/m^2. Efforts are being made to improve this performance and reduce the construction and operating costs of SMFC. To date, most SMFCs have been operated at a laboratory scale. In the future, scaling-up of SMFCs will be required to overcome the many hurdles and tackle the many new challenges. The objective of this study is to investigate the different aspects of optimal design of SMFC, which will be practised at field level.

Keywords Anode · Cathode · Electrotrophs · Exoelectrogens
Sediment

1 Introduction

Electricity production through a microbial fuel cell (MFC) was first observed by Potter in 1911 [1]. In the early 1990s, the researchers used different chemicals such as thonin, natural red, methyl viologen, potassium, ferricyanide,

S. Z. Abbas · M. Rafatullah (✉)
Division of Environmental Technology, School of Industrial Technology, Universiti Sains Malaysia, 11800 Penang, Malaysia
e-mail: mohd_rafatullah@yahoo.co.in; mrafatullah@usm.my

N. Horan et al. (eds.), *Anaerobic Digestion Processes*,
Green Energy and Technology, https://doi.org/10.1007/978-981-10-8129-3_3

anthraquinone-2,6-disulfonate and others for the exogenous transfer of electrons, but these are not eco-friendly in nature. Kim et al. [2] reported exogenous electron transfer without using such harmful compounds. Later Remiers et al. [3] found that marine sediment could be used as an MFC substrate for power generation and removal of contaminants. The first laboratory use of a sediment microbial fuel cell (SMFC) was reported by Bond et al. [4] who observed that acetate could be converted to CO_2 by direct electron transfer to the electrodes.

Presently for all available bioconversion techniques, the cost of primary biomass is very high to fulfil the required energy level. There are many ways to convert biomass to bioenergy as well as bioremediation of pollutants. During the 70s, methanogenic-based anaerobic digestion emerged rapidly and is now well established now. For biomass-to-bioenergy conversion, the hydrogen and ethanol fermentation is also practised [5]. Recently, the SMFC system has been established as a unique biotechnology to generate power from sediment and simultaneously remove the pollutants. The SMFCs generate electricity from sediment organic compounds directly, without gas production and combustion. Conversion can happen at low substrate concentrations and temperatures below 20 °C where anaerobic digestion is highly disadvantaged due to high solubility of the methane and low reaction rates. However, controversy indicates its applicability, the efficiency and apparently the future of the SMFC technology in the context of bioremediation and bioconversion. The performance of anaerobic SMFC totally depends upon the design of SMFC, in particular, the selection of SMFC material [6]. This study covers the optimal design of anaerobic SMFC, its operating mechanisms and future challenges to overcome its limitation for field application.

2 Design of Anaerobic SMFC

The basic components of an anaerobic SMFC are summarized in Table 1. The prototype of anaerobic SMFC consists of a graphite anode buried in the sediment and a cathode placed in the overlying water (Fig. 1). Oxygen is not supplied to the cathode chamber from outside, and also no permeable membrane is put between

Table 1 The basic components of anaerobic SMFC

Items	Material	Remarks
Cathode	Graphite, carbon cloth, Pt, RVC, carbon paper, graphite felt, Pt black	Necessary
Anode	Reticulated vitreous carbon (RVC), carbon paper, Pt, Pt black, graphite, graphite felt, carbon cloth	Necessary
Cathodic chamber	Polycarbonate, glass, Plexiglas	Optional
Anodic chamber	Plexiglas, glass, polycarbonate	Necessary

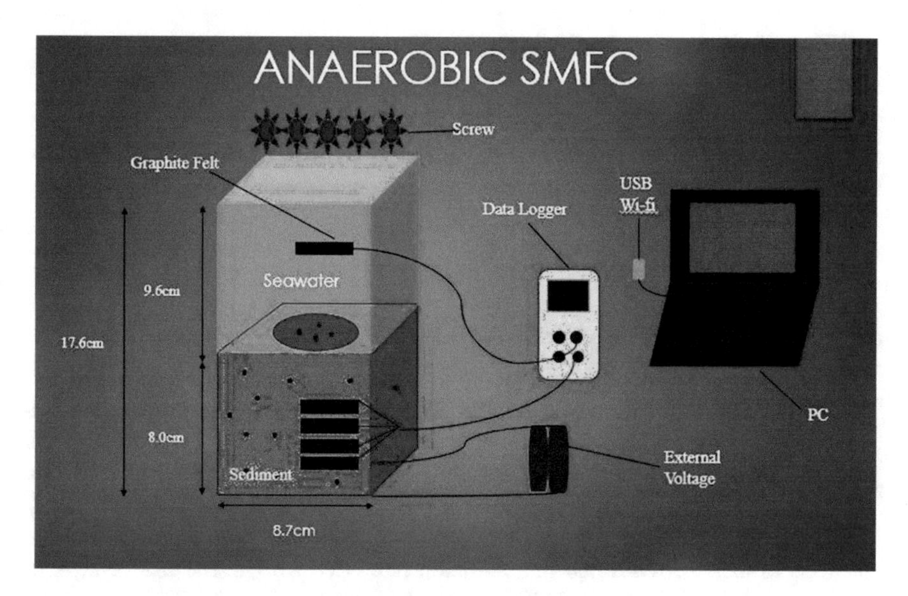

Fig. 1 Schematic model of anaerobic SMFC

these two chambers. The performance of anaerobic SMFCs depends upon the potential gradient between sediment and water surface.

Both anode and cathode are covered by coiled copper wire due to its high conduction properties. For efficient SMFC, the multi-anode system introduces in the anaerobic SMFC, about four graphite anodes connected with the single cathode. Both terminals are connected to the external circuit and monitored by a Wi-Fi multimeter data logger system. This model of anaerobic SMFC provides enough information about how to extract energy from the natural microbial consortia and can be used for the power generation and remediation of contaminants. A few reports are documented about sampling of remote sediments. The SMFC is normally dominated by natural microbes, but if the laboratory scale SMFC microbial strains are introduced into the field, they are normally dominated by wild strains due to competition and predation [7]. Anaerobic SMFC normally plays two roles during operational condition. (i) Anaerobic SMFC can produce power without extra field maintenance [8]. (ii) Anaerobic SMFC can generate electricity by electrochemical difference between overlying water and anaerobic sediment [9].

3 Types of Anaerobic SMFC

There are two main types of anaerobic SMFC. Those based on electricity production by microbes or by powering the microbes from an external source. In these SMFCs, methanogens are mostly responsible for all metabolic activities and produce the methane gas.

3.1 Anaerobic Non-stimulated SMFC

This is the simplest form of anaerobic SMFC without oxygen supply (Fig. 2). This SMFC is not stimulated by external current and normally more methane is produced than in an anaerobic stimulated SMFC. The only disadvantage of this SMFC is that the rate of bioremediation is much slower than anaerobic stimulated SMFC. In this SMFC, mostly exoelectrogenic microbes are present which transfer electrons to the electrodes and the oxidation rate is very high. The rate of biofilm formation is very high in this SMFC due to the large number of microbes that can act as exoelectrogens.

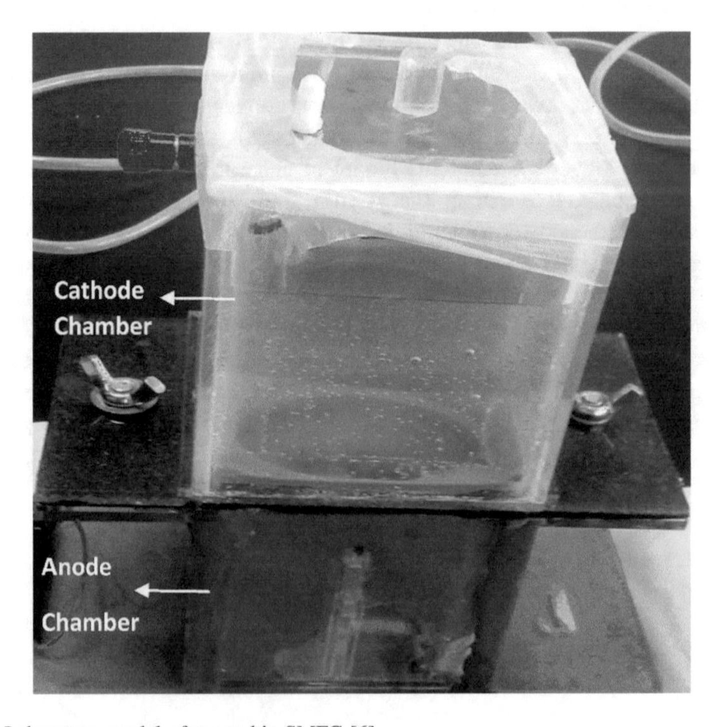

Fig. 2 Laboratory model of anaerobic SMFC [6]

3.2 Anaerobic Stimulated SMFC

The only difference with this SMFC is that it is stimulated by external current and this type of SMFC is highly suitable for electrotrophs. In this SMFC, the reduction/oxidation rate of organic and inorganic compounds is very high. Its disadvantage is the need to supply the power from the outside source but this problem can be solved either powering by solar cell or by stored energy. In this SMFC, mostly electrotrophs are present which accept electrons from the electrodes and are involved in the bioremediation of contaminants due to high rate of reduction. To date, only some species of *Geobacter* and *Shewanella* have been identified as electrotrophs, which is why the rate of biofilm formation in this SMFC is very low.

4 Substrate Used in the Anaerobic SMFCs

In the cathode chamber of SMFC, different substrates can be used. The power generation depends upon the composition of cathode substrates. The range of substrates varies from pure compounds to mixtures. In the last century, the main treatment process has been activated sludge but it requires a large energy input [10]. The addition of a second treatment process changes the status of contaminants that might be used for energy production. Table 2 enlists the substrates with different power generation capacities. It is very difficult to find out the exact performance of SMFC due to the different operating conditions, SMFC design, various microbes and different units that are used for measuring the current. The most common unit used to measure power generation is current generation per unit area of cathode and anode (mA/cm^2) or current generated per unit volume of the cell (mA/m^3) [11].

Acetate: in the most of SMFC, acetate is preferred instead of different types of wastewater. Acetate is a simple compound and easily initiates the electroactive pathways inside microbes. Acetate can be easily converted into different metabolic pathways such as anaerobic fermentation. In addition, acetate is the end product of many metabolic routes such as glycogenesis and the Entner–Doudoroff pathway [28].

Glucose: It is another substrate used in the SMFC. Glucose can run SMFC for a short time compared to other substrates. Rabaey et al. [29] found that from the glucose added about 216 W/m^3 power density can be obtained. The power generation of anaerobic sludge was compared with glucose. Anaerobic sludge produced about 0.3 mW/m^2 in the baffle SMFC and in the same system glucose produced about 161 mW/m^2 [30]. The energy conversion rate of glucose and acetate was also compared. The energy conversion rate of acetate was about 42% but only 3% with glucose. Glucose is more easily broken down than acetate and can be used in many competitive pathways like fermentation and methanogenesis.

Wastewater: Chemical and synthetic wastewaters are used in many SMFCs to control the operating conditions like conductivity and pH. Many growth media are

Table 2 Substrates used in sediment microbial fuel cells (SMFCs) and the maximum power generation

Type of substrate	Current density (mA/cm^2) at maximum power	Concentration	References
Acetate	0.8	1 g/L	[12]
Cellulose particles	0.02	4 g/L	[13]
Furfural	0.17	6.8 mM	[14]
Glucose	0.70	6.7 mM	[15]
Lactate	0.005	18 mM	[16]
Mannitol	0.58	1220 mg/L	[17]
Phenol	0.1	400 mg/L	[18]
Sodium formate	0.22	20 mM	[19]
Sorbitol	0.62	1220 mg/L	[17]
Sucrose	0.19	2674 mg/L	[20]
Xylose	0.74	6.7 mM	[15]
Brewery wastewater	0.2	2240 mg/L	[21]
Domestic wastewater	0.06	600 mg/L	[22]
Meat processing wastewater	0.115	1420 mg/L	[23]
Swine wastewater	0.015	8320 mg/L	[24]
Synthetic wastewater	0.008	510 mg/L	[25]
Xylose and humic acid	0.06	10 mM	[26]
Landfill leachate	0.0004	6000 mg/L	[27]

used for microbial growth like cysteine rich wastewater and wastewater with less sulphur. Many researchers used different synthetic wastewaters and reported different SMFC performance rates [31]. Brewery wastewater is ideal as an SMFC substrate because it is nd rich in carbohydrate with a reduced ammonium concentration. Starch processing wastewater contains a high percentage of starch, protein and carbohydrates and can be easily converted into many useful products in SMFC [18]. The azo dye-containing wastewater is a common effluent from textile industries. The colour of azo dye is very harmful to the aquatic ecosystem. Some compounds in the azo dye wastewater are more toxic. So the azo dye-containing wastewater was recently used as SMFC substrate to remove the colour and power generation [32].

5 Mechanisms of SMFC Electricity Generation

Many microbes are able to transfer the electrons to the electrodes via several routes (Fig. 3). Five groups of *Proteobacteria*, fungi, yeast, *Acidobacteria* and microalgae have shown the capacity for power generation. Four types of mechanisms exist in

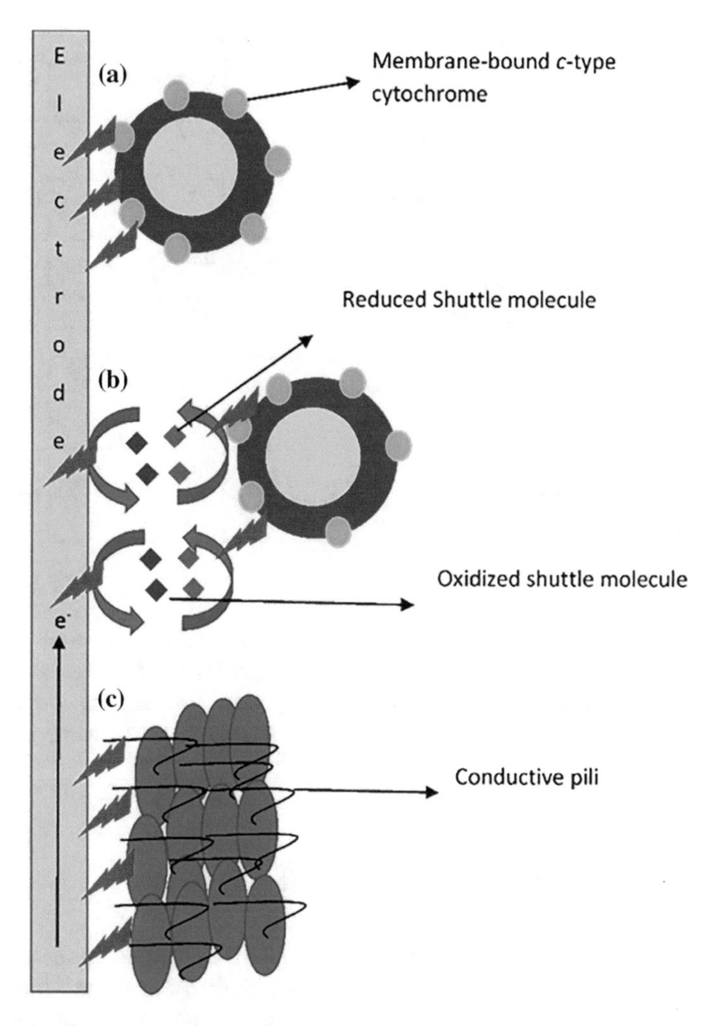

Fig. 3 **a** Short-range electron transfer, **b** electron transfer via redox-active proteins and **c** long-range electron transfer through conductive pili

the exoelectrogens. Table 3 shows the different exoelectrogens with their ways of electrons transfer respect to power generation.

5.1 Short-Range Electron Transfer

Bond et al. [4] first reported the electron transfer to the electrodes through self-produced electron shuttles molecules in the *Geobacter fermentans*. All Gram-positive and Gram-negative bacteria have the ability to transfer the electron

Table 3 Types of electron transfer mediators in different exoelectrogens

Exoelectrogens	Electron transfer intermediates	Current density (W/m^3)	References
Leptothrix discophora SP-6	Mn(IV) and Mn(II)	200	[36]
Acinetobacter calcoaceticus	Pyrroloquinoline quinone	1.3	[37]
Shewanella putrefaciens	Unidentified outer membrane bound redox compounds	0.7	[37]
Dechloromonas sp.	2,6-Anthraquinone disulfonate	0.6	[38]
Geobacter lovleyi	Methyl viologen	0.05	[39]
Desulfovibrio vulgaris	Methyl viologen	0.30	[40]
Chlorella vulgaris	Methylene blue	0.3	[41]
Klebsiella pneumoniae	2,6-di-tert-butyl-p-benzoquinone	199	[42]

to the electrodes through self-produced shuttle molecules. The most common microbial families able to self-produce electron shuttles are *Geobacteraceae* and *Desulfuromonadaceae*. The most common electron shuttles are c-type cytochromes. The c-type cytochromes include OmcA, MtrC, MtrE, MtrF and MtrD [33].

5.2 Electron Transfer Via Redox-Active Protein

Geobacter sulfurreducens has the ability to transfer the electrons to the electrodes through its anaerobic enzymatic metabolism. *G. sulfurreducens* can transfer the electrons to the different acceptors like fumarate and elemental sulphur. Many redox-active proteins are present in exoelectrogens like OmcS, OmcB, OmcT, OmcE and OmcZ [34].

5.3 Long-Range Electron Transfer Through Conductive Pili

Short, filamentous projections on a bacterial cell, used not for motility but for adhering to surface are called pili. These pili are involved in the transfer of electrons from exoelectrogens biofilm. These pili translated by *pilA* gene. These pili act as nanowires to transfer the electrons to the electrodes. These pili mostly present in the strains of *Methanothermobacter thermautotrophicus*, *Pelotomaculum thermopropionicum*, *G. sulfurreducens* and *Shewanella oneidensis* [35].

5.4 Direct Interspecies Electron Transfer

Previously, this mechanism only studied in the *G. sulfurreducens* and *Geobacter metallireducens*. In this mechanism, the strains transfer the electrons through each other and promote the mutual growth. Recently, this mechanism has been studied among aerobic *Synechocystis* and anaerobic *P. thermopropionicum* [43].

6 Pollutants Removal

The anaerobic SMFC can remove both organic and inorganic pollutants.

6.1 Removal of Organic Pollutants

The anaerobic SMFC is used to remove organic pollutants by generating electrochemical potential. The removal procedure consists of the oxidation of organic compounds in the anode chamber coupled with the oxidation of O_2 in the cathode chamber. Many hydrocarbons like nitrogen- and chlorine-containing compounds are reported to be removed by SMFC [44]. The wastewater contains many organic compounds like propionate, acetate and butyrate that can be completely metabolized to H_2O and CO_2. A combination of anodo-philes and cathode-philes microorganisms is more suitable to degrade the wide range of organic compounds. The SMFC microbe's growth rapidly during organic removal due to suitable conditions. The food processing wastewater, sanitary wastes, corn stover and swine wastewater contain a high amount of organic compounds that can be removed by SMFC and in some cases, about 80% COD removal rate was reported [45]. Sherafatmand and Ng [46] removed the naphthalene, acenaphthene and phenanthrene 41.7, 52.5 and 36.8%, respectively. Pous et al. [47] removed NO_3^- and NO_2^- 12.14 ± 3.59 and 0.14 ± 0.13 mg, respectively.

Nitrogen normally is removed by conventional biological methods such as nitrification and denitrification. But in ANAMMOX, the ammonia could also be anaerobically oxidized by using nitrite as electron acceptor. Recently, an ANAMMOX-like process was reported in the anaerobic SMFC in which ammonia was oxidized anaerobically. Conventionally nitrogen accepts the electrons from organic compounds and is converted into nitrogen gas. In the anaerobic SMFC, the nitrate can be used as terminal electron acceptor and reduced to produce the electricity with positive electric potential Eqs. (1) and (2) [48].

$$CH_3COO + 2H_2O \rightarrow 2CO_2 + 7H + 8e^- \quad E_0 = -0.28 \text{ V versus SHE} \qquad (1)$$

$$2NO_3 + 10e + 12H^+ \rightarrow N_2 + 6H_2O \quad E_0 = +0.70 \text{ V versus SHE} \tag{2}$$

Unlike the conventional denitrification process dependent upon the heterotrophic exoelectrogens in the anaerobic SMFC, denitrification is carried out by autotrophic electrophiles that accept the electrons from electrode and perform denitrification. This kind of phenomena was first reported by Gregory et al. [49] in *G. metallireducens* which can accept the electrons from the electrodes and reduce nitrate to nitrite. This electrophilic mechanism was also reported in the mixed culture of exoelectrogens to reduce nitrate to nitrogen gas. These autotrophs microbes use hydrogen as electron donor which generate in the cathode chamber. These autotrophic denitrifying microbes are known as autohydrogenotrophs.

Nitrate is reduced to nitrite (NO_2^-) which further combines with hydrogen to produce nitric oxide (NO). Then, this compound is reduced to nitrous oxide (N_2O)

Table 4 Mechanism of denitrification as electron donor (hydrogen gas)

Reaction steps	Process
$2H_2O + 2e^- \rightarrow H_2 + 2OH^-$	Electrolysis of water at cathode
$NO_3^- + H_2 \rightarrow NO_2^- + H_2O$	Reduction of nitrate
$NO_2^- + H^+ + 0.5H_2 \rightarrow NO + H_2O$	Reduction of nitrate
$2NO + H_2 \rightarrow N_2O + H_2O$	Reduction of nitric oxide
$N_2O + H_2 \rightarrow N_2 + H_2O$	Reduction of nitric oxide
$2NO_3^- + 5H_2 + 2H^+ \rightarrow N_2 + 6H_2O$	Overall denitrification

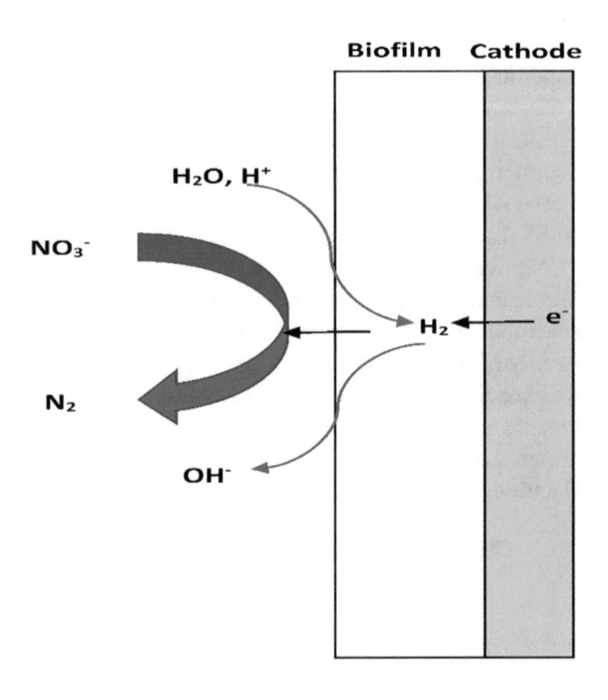

Fig. 4 Schematic denitrification reaction

and finally forms nitrogen gas (N_2) [50]. The reaction mechanism is given in Table 4, and the schematic reaction is shown in Fig. 4.

6.2 Removal of Inorganic Pollutants

The SMFC can also remove inorganic pollutants. Phosphorus is an inorganic compound and can be removed by chemical and biological process but the biological process is most favourite due to its low cost. In the anaerobic SMFC, the microbes normally accumulate phosphorus intracellularly higher than the requirement of their normal growth, and then, it is finally removed in the waste sludge.

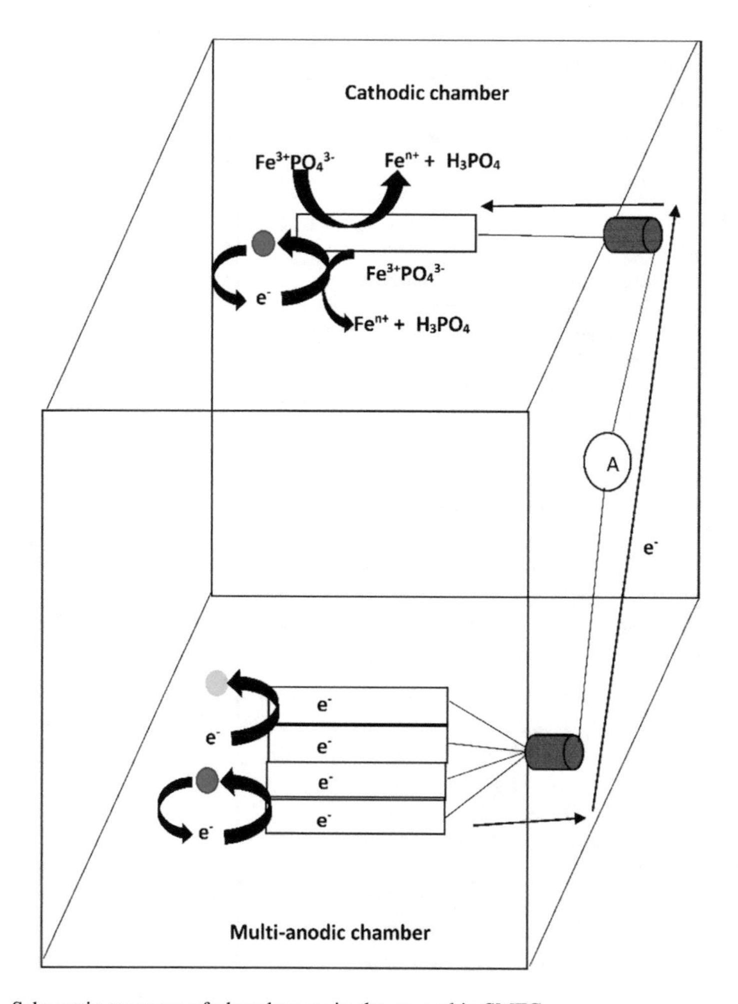

Fig. 5 Schematic recovery of phosphorous in the anaerobic SMFC

Phosphorus has not been studied as extensively in SMFC as nitrogen-containing compounds but in Fig. 5, the recovery of phosphorus has been shown.

The anaerobic SMFC can also remove toxic heavy metals like Cu (II), Zn (II), U (VI), Cr (VI) and Cd (II) to insoluble and less toxic forms. Many SMFC electrophiles reduce heavy metals into insoluble or less toxic forms which release into the overlying water or absorb on the electrodes surface.

Abbas et al. [51] removed arsenic (46.56%), cadmium (60.75%) and lead (42.49%) in the non-aerated SMFC. Li et al. [52] also removed about 66.2% of Cr through the reduction of $Cr_2O_7^{2-}$ to Cr_2O_3 in the anaerobic SMFC. Not all the heavy metals can be remediated by reduction. Some heavy metals require reduction

Table 5 List of microbes with type of pollutants removal

Microbes	Contaminates remediation	References
Saccharomyces cerevisiae	Zn^{2+}	[54]
Saccharomyces rimosus *Penicillium chrysogenum*	Cu^{2+}	[55]
Phanerochaete chrysosporium *Streptomyces clavulgerus*	Pb^{2+}	[56]
Bacillus lentus	$Cd2^{+}$	[57]
Candida sp.	Cu^{2+}	[57]
Pseudomonas aeruginosa, Cellulomonas turbata *Shewanella oneidensis, Aspergillus parasiticus* *Corynebacterium hoagie, Bacillus megaterium* *Bacillus maroccanus, Pseudomonas* sp., *Trichococcus pasteurii, Aspergillus niger*	Cr(VI)	[58]
Bacillus sp., *Desulfovibrio desulfuricans*	Se(VI)	[59]
Desulfovibrio desulfuricans, Bacillus subtilis, Rhodospirillum rubrum, Microbacterium arborescens, Pseudomonas fluorescens, Anaeromyxobacter dehalogenans	Se(IV)	[60]
Geobacillus sp., *Verticillium luteoalbum, Shewanella algae, Cupriavidus metallidurans*	Au(III)	[61]
Cupriavidus metallidurans, Escherichia coli, Cupriavidus necator, Geobacter sulfurreducens	Pd(II)	[62]
Shewanella oneidensis	V(V)	[63]
Plectonema boryanum	Ag(I)	[64]
Desulfotomaculum auripigmentum, Chrysiogenes arsenates	As(V)	[65]
Shewanella putrefaciens, Bacillus subterraneus	Mn(IV)	[66]
Bacillus subterraneus, Geobacter psychrophilus, Shewanella putrefaciens, Geobacter bemidjiensis, Bacillus subterraneus	Fe(III)	[67]
Klebsiella sp., *Serratia* sp., *Acinetobacter calcoaceticus*	Mo(VI)	[68]
Bacillus cereus, Anoxybacillus sp., *Shewanella oneidensis*	Hg(II)	[69]
Escherichia coli, Geobacter sulfurreducens, Desulfovibrio desulfuricans	Tc(VII)	[70]

to convert into less toxic forms like arsenic which needs oxidation with ferric oxides. Under reduction, mercury converts to the more toxic methyl mercury. So oxidation is required to pump out the mercury from the cells. The different pollutants removal by different microbes is shown in Table 5.

7 Mechanisms of Pollutants Removal

The microorganisms that act as electron acceptors from cathode donor are called electrotrophs. As shown in Fig. 6, many limitations of electron transfer have been overcome by exoelectrogens. Many studies reported that the gene expression of electrotrophs is totally different from exoelectrogens like pilA and omcZ genes [53].

But in the case of fumarate reduction by *G. sulfurreducens*, the electrons enter into bacteria through same pathways as electron moving from bacteria to electrodes. *G.*

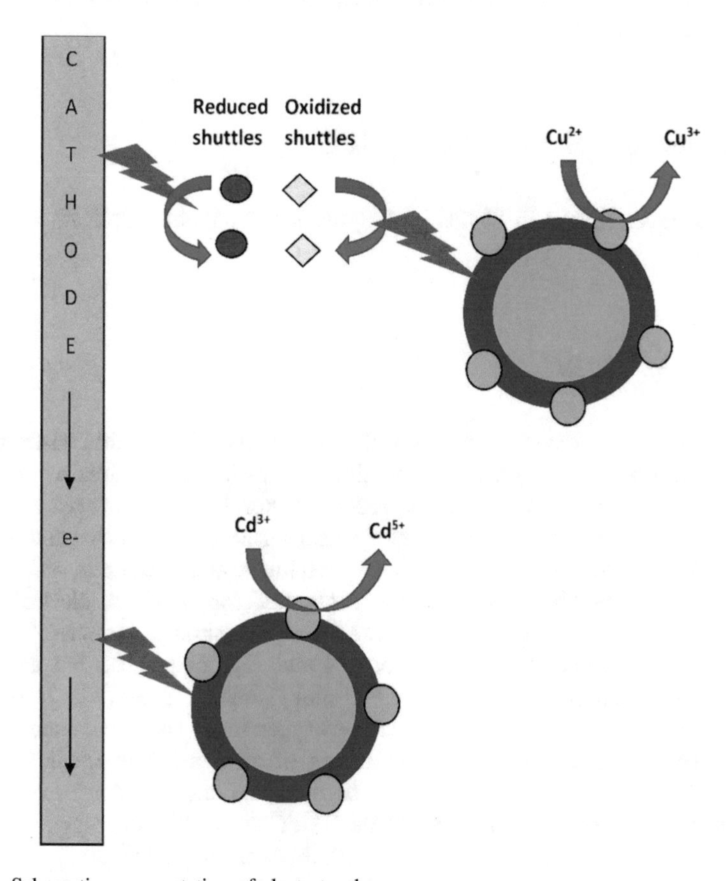

Fig. 6 Schematic representation of electrotrophs

sulfurreducens reduces soluble U(VI) to insoluble U(IV) and also soluble, toxic Cr (VI) to insoluble, less toxic Cr(III) by accepting the electrons from the electrodes [71].

In this process, energy is supplied by electrodes to produce an electron potential that makes favourable conditions for oxidation and reduction and if CO_2 is fixed, then this process is called electrosynthesis, similar to photosynthesis. Initially, the cathode was considered to act as electron donor to pure cultures of *Geobacter* spp. Recently, more species have been identified that may be electrotrophs in the pure culture of biofilm like *Sporomusa ovata*, *G. sulfurreducens*, *Sporomusa sphaeroides*, *Clostridium ljungdahlii*, *Clostridium aceticum*, *Moorella thermoacetica* and *G. metallireducens* [72]. These bacteria are mostly present in the single layer on the cathode but the cathode also attracts the mixed culture of electrotrophs with the ability of biosynthesis, improving bioremediation and biosensing. Normally, a thick biofilm is observed on the anode and thin biofilms on the cathode with pure culture. Recently, thick biofilms of mixed culture were observed on the cathode that will be helpful for new investigations. The biofilm dynamics and products totally depend upon the amount of current given to the cathode. This is very helpful to understanding the nature of interspecies interactions for the electron transfer.

8 Application of Anaerobic SMFC

The recent investigation in SMFC scaling-up, many possible applications can be considered.

8.1 Bio-hydrogen

Under the normal operating conditions, the production of bio-hydrogen from protons and electrons is theoretically impossible. Normally, the protons move to the cathode and combine with oxygen and produce water. For bio-hydrogen synthesis, there is need to increase cathode potential to overcome the thermodynamic barrier. In this process, electrons and protons are combined at the cathode and formed hydrogen. Theoretically, about 110 mV is required for the direct electrolysis of water at pH 7.0 because some energy arises from oxidation of biomass. The conventional fermentation can produce about 4 mol H_2/mol glucose but anaerobic SMFCs can potentially generate about 8–9 mol H_2/mol glucose [73]. In the production of bio-hydrogen, there is no need for oxygen in the cathode chamber so the SMFC efficiency can be increased by leaking of oxygen to the anode from the cathode. The produced hydrogen can be accumulated and stored for later usage. This hydrogen can be used for powering the SMFC during a low energy production stage. So this is a very good source of renewable energy production.

8.2 Biosensor

The SMFC can be used as pollutants biosensor and for in situ monitoring and control. The directly proportional relation between strength of wastewater and Coulombic efficiency make the SMFC suitable to be used as BOD biosensor. The accurate BOD value normally measured by Coulombic efficiency. Many researchers worked on the calculation of Coulombic efficiency by using the strength of wastewater [74, 75]. However, a higher BOD concentration needs a long response because the Coulombic efficiency can be calculated after the BOD has depleted in the wastewater. Many efforts are made for accurate measuring of BOD. Low BOD values can be monitored measuring the amount of current passing through SMFC. At high BOD, the amount of current increased in an oligotroph-type SMFC. At this stage, due to the high substrate concentration, the anodic reactions are very slow. This type of biosensing can be used for the real-time BOD calculation of secondary effluents, surface water and highly diluted BOD samples. This biosensor has many advantages over other BOD sensor as SMFC has good operational properties, accuracy and reproducibility. The SMFC biosensor can be operated for 5 years without any maintenance [6].

8.3 Bioelectronics

The exoelectrogens can be used for the detection of different chemical and pollutants. The produced signal can be translated into electrical peaks, which lead to the development of biological and biocomputing sensors. The high conduction of exoelectrogens biofilms and pili lead to the development of biodevices. There are many advantages of bioelectronics: (i) charge can be transmitted and stored underground water, (ii) devices can be constructed from less toxic material, (iii) bioelectronics have the ability of self-repairing and replication, and (iv) the high accumulation of c-type cytochromes and conduction of *G. sulfurreducens* make it as super-capacitors. Further investigations are required about the metal like conduction of these nanowires.

9 Future Directions

For the efficient power generation and bioremediation of pollutants, more investigations are needed.

9.1 Modelling

On the basis of biology, geochemistry and electrochemistry, the optimal SMFC model can be predicted. Others factors like depth of water, distance between electrodes and anode depth are needed to investigate prior to field implementation. As the electrodes distance increases, the ohmic losses also increase. So a floating SMFC may be future solution.

9.2 Monitoring

The next main challenge is to monitor the interaction of electrodes and microorganisms. So more controlled SMFC model can be constructed. Molecular techniques like in situ hybridization and metagenomics can be used for better control monitoring of exoelectrogens.

9.3 Selection of Electrode Material

The composition of electrodes material is very important because the electrodes act as electrons acceptor and donor. So the electrodes material should be resistant to biodegradation/corrosion, high surface area, mechanically durable, high porosity and cost-effective. The Fe^{3+} woven and Mn(IV) woven material may be future electrodes material solution because this material has property of synergistic interaction with exoelectrogens.

9.4 Electromicrobiology

Many future research directions in SMFC microbiology are needed such as interactions of pure and mix microbial cultures with electrodes, conduction of pili, biofilm formation and other conductive exoelectrogens.

9.5 Overlying Water

The pH, temperature, BOD and COD are very important for the efficient performance of SMFC. The dissolved oxygen is very important for current generation. More investigation is needed to find the technical solution to improve the oxygen concentration into overlying water.

9.6 SMFC Scaling-up

The production of power at a Watt level is not reported in any SMFC study. So there is need to design an optimal SMFC model for practise at field level. So further investigations are needed to optimize the following factors like reactor design, operational mode, material selection and environmental sustainability of system.

10 Conclusions

This study has summarized the model of anaerobic SMFC and its operating mechanisms that could be possibly used for the power generation and bioremediation of pollutants. From the current perspectives, the selection of electrodes material and its cost and efficient performance will be future attractive directions for SMFC. The integration of SMFC with wastewater treatment technologies will offer more promising tools for wastewater treatment. Future directions and challenges for the SMFC scaling-up are also discussed. These challenges will be overcome by joint accomplishment from different fields like chemistry, geology, biology, physics, biotechnology and computer science.

Acknowledgements The authors would like to express their appreciation to Universiti Sains Malaysia Global Fellowship (USMGF) and RUI grant (1001/PTEKIND/8011044) for the support and research facilities for this study.

References

1. Potter MC (1911) Electrical effects accompanying the decomposition of organic compounds. Proc R Soc Lond Ser B, Containing Papers Biol Character 84(571):260–276
2. Kim BH, Ikeda T, Park HS, Kim HJ, Hyun MS, Kano K, Takagi K, Tatsumi H (1999) Electrochemical activity of an Fe (III)-reducing bacterium, *Shewanella putrefaciens* IR-1, in the presence of alternative electron acceptors. Biotechnol Tech 13(7):475–478
3. Reimers CE, Tender LM, Fertig S, Wang W (2001) Harvesting energy from the marine sediment–water interface. Environ Sci Technol 35(1):192–195
4. Bond DR, Holmes DE, Tender LM, Lovley DR (2002) Electrode-reducing microorganisms that harvest energy from marine sediments. Science 295(5554):483–485
5. Pham T, Rabaey K, Aelterman P, Clauwaert P, De Schamphelaire L, Boon N, Verstraete W (2006) Microbial fuel cells in relation to conventional anaerobic digestion technology. Eng Life Sci 6(3):285–292
6. Abbas SZ, Rafatullah M, Ismail N, Syakir MI (2017) A review on sediment microbial fuel cells as a new source of sustainable energy and heavy metal remediation: mechanisms and future prospective. Int J Energy Res 41(9):1242–1264
7. Donovan C, Dewan A, Heo D, Beyenal H (2008) Batteryless, wireless sensor powered by a sediment microbial fuel cell. Environ Sci Technol 42(22):8591–8596

8. ElMekawy A, Hegab HM, Vanbroekhoven K, Pant D (2014) Techno-productive potential of photosynthetic microbial fuel cells through different configurations. Renew Sustain Energy Rev 39:617–627
9. Kothapalli AL (2013) Sediment microbial fuel cell as sustainable power resource. The University of Wisconsin-Milwaukee
10. Macknick J, Newmark R, Heath G, Hallett K (2012) Operational water consumption and withdrawal factors for electricity generating technologies: a review of existing literature. Environ Res Lett 7(4):1–10
11. Pant D, Van Bogaert G, Diels L, Vanbroekhoven K (2010) A review of the substrates used in microbial fuel cells (MFCs) for sustainable energy production. Bioresour Technol 101 (6):1533–1543
12. Logan BE (2009) Exoelectrogenic bacteria that power microbial fuel cells. Nat Rev Microbiol 7(5):375–381
13. Rezaei F, Richard TL, Logan BE (2009) Analysis of chitin particle size on maximum power generation, power longevity, and Coulombic efficiency in solid–substrate microbial fuel cells. J Power Sources 192(2):304–309
14. Luo Y, Liu G, Zhang R, Zhang C (2010) Power generation from furfural using the microbial fuel cell. J Power Sources 195(1):190–194
15. Catal T, Li K, Bermek H, Liu H (2008) Electricity production from twelve monosaccharides using microbial fuel cells. J Power Sources 175(1):196–200
16. Manohar AK, Mansfeld F (2009) The internal resistance of a microbial fuel cell and its dependence on cell design and operating conditions. Electrochim Acta 54(6):1664–1670
17. Catal T, Xu S, Li K, Bermek H, Liu H (2008) Electricity generation from polyalcohols in single-chamber microbial fuel cells. Biosens Bioelectron 24(4):849–854
18. Luo H, Liu G, Zhang R, Jin S (2009) Phenol degradation in microbial fuel cells. Chem Eng J 147(2):259–264
19. Ha PT, Tae B, Chang IS (2007) Performance and bacterial consortium of microbial fuel cell fed with formate. Energy Fuels 22(1):164–168
20. Behera M, Ghangrekar M (2009) Performance of microbial fuel cell in response to change in sludge loading rate at different anodic feed pH. Bioresour Technol 100(21):5114–5121
21. Feng Y, Wang X, Logan BE, Lee H (2008) Brewery wastewater treatment using air-cathode microbial fuel cells. Appl Microbiol Biotechnol 78(5):873–880
22. Wang X, Feng Y, Ren N, Wang H, Lee H, Li N, Zhao Q (2009) Accelerated start-up of two-chambered microbial fuel cells: effect of anodic positive poised potential. Electrochim Acta 54(3):1109–1114
23. Heilmann J, Logan BE (2006) Production of electricity from proteins using a microbial fuel cell. Water Environ Res 78(5):531–537
24. Min B, Kim J, Oh S, Regan JM, Logan BE (2005) Electricity generation from swine wastewater using microbial fuel cells. Water Res 39(20):4961–4968
25. Mohan SV, Mohanakrishna G, Reddy BP, Saravanan R, Sarma P (2008) Bioelectricity generation from chemical wastewater treatment in mediatorless (anode) microbial fuel cell (MFC) using selectively enriched hydrogen producing mixed culture under acidophilic microenvironment. Biochem Eng J 39(1):121–130
26. Huang L, Angelidaki I (2008) Effect of humic acids on electricity generation integrated with xylose degradation in microbial fuel cells. Biotechnol Bioeng 100(3):413–422
27. Greenman J, Gálvez A, Giusti L, Ieropoulos I (2009) Electricity from landfill leachate using microbial fuel cells: comparison with a biological aerated filter. Enzyme Microb Technol 44 (2):112–119
28. Biffinger JC, Byrd JN, Dudley BL, Ringeisen BR (2008) Oxygen exposure promotes fuel diversity for *Shewanella oneidensis* microbial fuel cells. Biosens Bioelectron 23(6):820–826
29. Rabaey K, Lissens G, Siciliano SD, Verstraete W (2003) A microbial fuel cell capable of converting glucose to electricity at high rate and efficiency. Biotechnol Lett 25(18):1531–1535

30. Chae K-J, Choi M-J, Lee J-W, Kim K-Y, Kim IS (2009) Effect of different substrates on the performance, bacterial diversity, and bacterial viability in microbial fuel cells. Bioresour Technol 100(14):3518–3525
31. Rodrigo MA, Cañizares P, García H, Linares JJ, Lobato J (2009) Study of the acclimation stage and of the effect of the biodegradability on the performance of a microbial fuel cell. Bioresour Technol 100(20):4704–4710
32. Sun J, Y-y Hu, Bi Z, Y-q Cao (2009) Simultaneous decolorization of azo dye and bioelectricity generation using a microfiltration membrane air-cathode single-chamber microbial fuel cell. Bioresour Technol 100(13):3185–3192
33. Wei T, Ma H, Nakano A (2016) Decaheme cytochrome MtrF adsorption and electron transfer on gold surface. J Phys Chem Lett 7(5):929–936
34. Estevez-Canales M, Kuzume A, Borjas Z, Füeg M, Lovley D, Wandlowski T, Esteve-Núñez A (2015) A severe reduction in the cytochrome C content of *Geobacter sulfurreducens* eliminates its capacity for extracellular electron transfer. Environ Microbiol Rep 7(2):219–226
35. Lovley DR (2011) Live wires: direct extracellular electron exchange for bioenergy and the bioremediation of energy-related contamination. Energy Environ Sci 4(12):4896–4906
36. Nguyen TA, Lu Y, Yang X, Shi X (2007) Carbon and steel surfaces modified by *Leptothrix discophora* SP-6: characterization and implications. Environ Sci Technol 41(23):7987–7996
37. Freguia S, Rabaey K, Yuan Z, Keller J (2008) Sequential anode–cathode configuration improves cathodic oxygen reduction and effluent quality of microbial fuel cells. Water Res 42 (6):1387–1396
38. Lefebvre O, Al-Mamun A, Ng H (2008) A microbial fuel cell equipped with a biocathode for organic removal and denitrification. Water Sci Technol 58(4):881–885
39. Strycharz SM, Woodard TL, Johnson JP, Nevin KP, Sanford RA, Löffler FE, Lovley DR (2008) Graphite electrode as a sole electron donor for reductive dechlorination of tetrachlorethene by *Geobacter lovleyi*. Appl Environ Microbiol 74(19):5943–5947
40. Lojou E, Durand M, Dolla A, Bianco P (2002) Hydrogenase activity control at *Desulfovibrio vulgaris* cell-coated carbon electrodes: biochemical and chemical factors influencing the mediated bioelectrocatalysis. Electroanalysis 14(13):913–922
41. Villano M, Aulenta F, Ciucci C, Ferri T, Giuliano A, Majone M (2010) Bioelectrochemical reduction of CO_2 to CH_4 via direct and indirect extracellular electron transfer by a hydrogenophilic methanogenic culture. Bioresour Technol 101(9):3085–3090
42. Logan BE (2005) Simultaneous wastewater treatment and biological electricity generation. Water Sci Technol 52(1–2):31–37
43. Shrestha PM, Rotaru A-E (2014) Plugging in or going wireless: strategies for interspecies electron transfer. Front Microbiol 5:1–8
44. Pandey P, Shinde VN, Deopurkar RL, Kale SP, Patil SA, Pant D (2016) Recent advances in the use of different substrates in microbial fuel cells toward wastewater treatment and simultaneous energy recovery. Appl Energy 168:706–723
45. Zhan G, Zhang L, Li D, Su W, Tao Y, Qian J (2012) Autotrophic nitrogen removal from ammonium at low applied voltage in a single-compartment microbial electrolysis cell. Bioresour Technol 116:271–277
46. Sherafatmand M, Ng HY (2015) Using sediment microbial fuel cells (SMFCs) for bioremediation of polycyclic aromatic hydrocarbons (PAHs). Bioresour Technol 195:122–130
47. Pous N, Puig S, Coma M, Balaguer MD, Colprim J (2013) Bioremediation of nitrate-polluted groundwater in a microbial fuel cell. J Chem Technol Biotechnol 88(9):1690–1696
48. Manassaram DM, Backer LC, Moll DM (2007) A review of nitrates in drinking water: maternal exposure and adverse reproductive and developmental outcomes. Ciencia & saude coletiva 12(1):153–163
49. Gregory KB, Bond DR, Lovley DR (2004) Graphite electrodes as electron donors for anaerobic respiration. Environ Microbiol 6(6):596–604

50. Mook W, Chakrabarti M, Aroua M, Khan G, Ali B, Islam M, Hassan MA (2012) Removal of total ammonia nitrogen (TAN), nitrate and total organic carbon (TOC) from aquaculture wastewater using electrochemical technology: a review. Desalination 285:1–13
51. Abbas SZ, Rafatullah M, Ismail N, Nastro RA (2017) Enhanced bioremediation of toxic metals and harvesting electricity through sediment microbial fuel cell. Int J Energy Res 41 (14):2345–2355
52. Li Z, Zhang X, Lei L (2008) Electricity production during the treatment of real electroplating wastewater containing Cr^{6+} using microbial fuel cell. Process Biochem 43(12):1352–1358
53. Lovley DR, Nevin KP (2013) Electrobiocommodities: powering microbial production of fuels and commodity chemicals from carbon dioxide with electricity. Curr Opin Biotechnol 24 (3):385–390
54. Cho DH, Yoo MH, Kim EY (2004) Biosorption of lead (Pb^{2+}) from aqueous solution by *Rhodotorula aurantiaca*. J Microbiol Biotechnol 14(2):250–255
55. Alkorta I, Hernández-Allica J, Becerril J, Amezaga I, Albizu I, Garbisu C (2004) Recent findings on the phytoremediation of soils contaminated with environmentally toxic heavy metals and metalloids such as zinc, cadmium, lead, and arsenic. Rev Environ Sci Biotechnol 3 (1):71–90
56. Nozaki K, Beh CH, Mizuno M, Isobe T, Shiroishi M, Kanda T, Amano Y (2008) Screening and investigation of dye decolorization activities of basidiomycetes. J Biosci Bioeng 105 (1):69–72
57. Ramalho PA (2005) Degradation of dyes with microorganisms: studies with ascomycete yeasts. Universidade do minho escola de ciências, escola de engenharia, Portugal
58. Shugaba A, Buba F, Kolo B, Nok A, Ameh D, Lori J (2012) Uptake and reduction of hexavalent chromium by *Aspergillus niger* and *Aspergillus parasiticus*. J Petrol Environ Biotechnol 3(3):1–8
59. Tandukar M, Huber SJ, Onodera T, Pavlostathis SG (2009) Biological chromium (VI) reduction in the cathode of a microbial fuel cell. Environ Sci Technol 43(21):8159–8165
60. He Q, Yao K (2011) Impact of alternative electron acceptors on selenium (IV) reduction by *Anaeromyxobacter dehalogenans*. Bioresour Technol 102(3):3578–3580
61. Correa-Llantén DN, Muñoz-Ibacache SA, Castro ME, Muñoz PA, Blamey JM (2013) Gold nanoparticles synthesized by *Geobacillus* sp. strain ID17 a thermophilic bacterium isolated from Deception Island, Antarctica. Microb Cell Fact 12(1):1–6
62. Yates MD, Cusick RD, Logan BE (2013) Extracellular palladium nanoparticle production using *Geobacter sulfurreducens*. Acs Sustain Chem Eng 1(9):1165–1171
63. Deplanche K, Merroun ML, Casadesus M, Tran DT, Mikheenko IP, Bennett JA, Zhu J, Jones IP, Attard GA, Wood J (2012) Microbial synthesis of core/shell gold/palladium nanoparticles for applications in green chemistry. J R Soc Interface 9(72):1705–1712
64. Lengke MF, Fleet ME, Southam G (2007) Biosynthesis of silver nanoparticles by filamentous cyanobacteria from a silver (I) nitrate complex. Langmuir 23(5):2694–2699
65. Gauthier D, Søbjerg LS, Jensen KM, Lindhardt AT, Bunge M, Finster K, Meyer RL, Skrydstrup T (2010) Environmentally benign recovery and reactivation of palladium from industrial waste by using gram-negative bacteria. ChemSusChem 3(9):1036–1039
66. Shukor M, Rahman M, Suhaili Z, Shamaan N, Syed M (2010) Hexavalent molybdenum reduction to Mo-blue by *Acinetobacter calcoaceticus*. Folia Microbiol 55(2):137–143
67. Nevin KP, Holmes DE, Woodard TL, Hinlein ES, Ostendorf DW, Lovley DR (2005) *Geobacter bemidjiensis* sp. nov. and *Geobacter psychrophilus* sp. nov., two novel Fe (III)-reducing subsurface isolates. Int J Syst Evol Microbiol 55(4):1667–1674
68. Lim H, Syed M, Shukor M (2012) Reduction of molybdate to molybdenum blue by *Klebsiella* sp. strain hkeem. J Basic Microbiol 52(3):296–305
69. Kritee K, Blum JD, Barkay T (2008) Mercury stable isotope fractionation during reduction of Hg (II) by different microbial pathways. Environ Sci Technol 42(24):9171–9177
70. Rahman M, Shukor M, Suhaili Z, Mustafa S, Shamaan N, Syed M (2009) Reduction of Mo (VI) by the bacterium *Serratia* sp. strain DRY5. J Environ Biol 30(1):65–72

71. Moscoviz R, De Fouchécour F, Santa-Catalina G, Bernet N, Trably E (2017) Cooperative growth of *Geobacter sulfurreducens* and *Clostridium pasteurianum* with subsequent metabolic shift in glycerol fermentation. Sci Rep. https://doi.org/10.1038/srep44334
72. Hartline RM, Call DF (2016) Substrate and electrode potential affect electrotrophic activity of inverted bioanodes. Bioelectrochemistry 110:13–18
73. Dumas C, Mollica A, Féron D, Basséguy R, Etcheverry L, Bergel A (2007) Marine microbial fuel cell: use of stainless steel electrodes as anode and cathode materials. Electrochim Acta 53 (2):468–473
74. He Z, Angenent LT (2006) Application of bacterial biocathodes in microbial fuel cells. Electroanalysis 18(19–20):2009–2015
75. Abbas S, Rafatullah M, Hossain K, Ismail N, Tajarudin H, Abdul Khalil HPS (2017) A review on mechanism and future perspectives of cadmium-resistant bacteria. Int J Environ Sci Technol. https://doi.org/10.1007/s13762-017-1400-5

A Current Review on the Application of Enzymes in Anaerobic Digestion

Mariani Rajin

Abstract Although the anaerobic digestion process is widely applied in waste management, it is recognised that the hydrolysis step in the treatment is a bottleneck that can restrict the rate that methane is produced. Enzyme addition during hydrolysis of a substrate has been reported as a promising alternative to overcome this limitation. This chapter presents a review of the supplementation of enzymes to facilitate the hydrolysis process of various types of substrates in the anaerobic digestion system.

Keywords Anaerobic digestion · Enzyme · Hydrolysis · Anaerobic digestion pretreatment

1 Introduction

Anaerobic digestion is a microbial process converting biodegradable organic material in the absence of oxygen into other forms of product such as biogas, anaerobic biomass and organic residuals [1]. This technology is widely applied in waste management including septic tanks, sludge digesters, agriculture residuals and energy crops, wastewater treatment, hazardous waste management and agricultural waste management [2, 3].

It is a versatile and effective technology. For instance, the anaerobic digester system is cost-effective in solving waste problems. Moreover, the application of this technology has a profound effect on reducing environmental impacts and enhancing the production of biogas. Owing to its benefits, anaerobic digestion has been discussed in numerous publications over the past decades.

Generally, anaerobic digestion process can be considered to occur in four sequential steps, namely, hydrolysis, acidogenesis, acetogenesis and methanogenesis

---

M. Rajin (✉)
Chemical Engineering Programme, Faculty of Engineering, Universiti Malaysia Sabah, Kota Kinabalu, Sabah, Malaysia
e-mail: mariani@ums.edu.my

N. Horan et al. (eds.), *Anaerobic Digestion Processes*,
Green Energy and Technology, https://doi.org/10.1007/978-981-10-8129-3_4

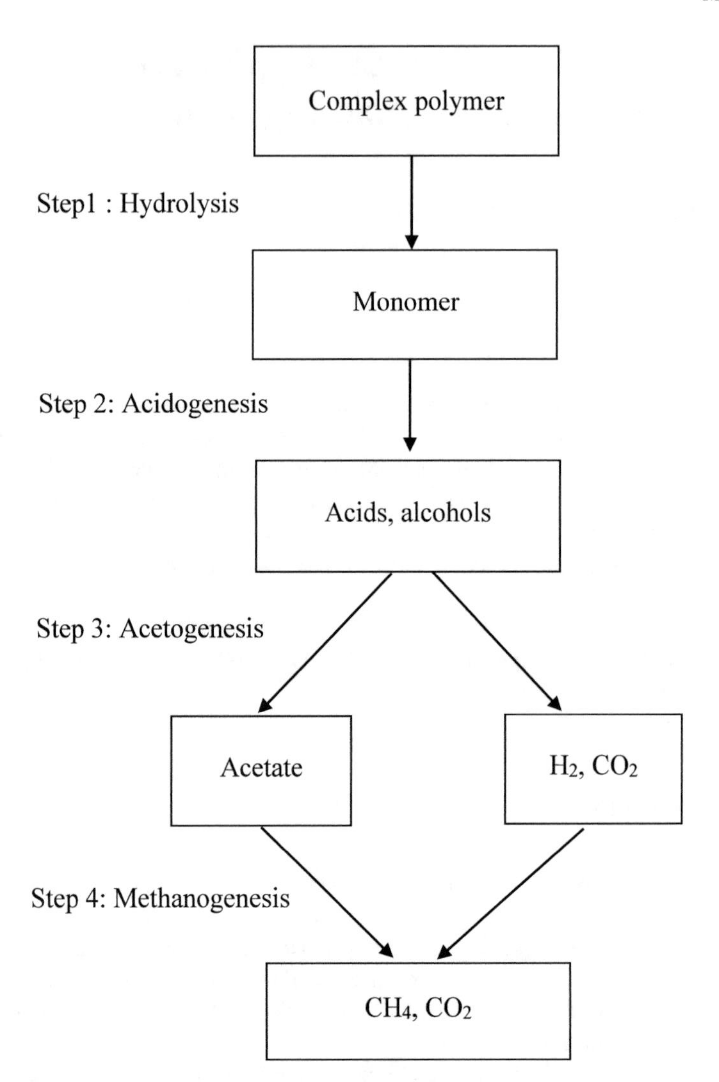

Fig. 1 Main steps in anaerobic digestion

(Fig. 1; Table 1). The technologies and operating mechanisms in each step have been well explained by previous researches [1, 4, 5]. Among these steps, hydrolysis is known as the rate-limiting step in anaerobic digestion.

During hydrolysis, complex polymers such as carbohydrates, lipids and proteins are converted into soluble monomers such as amino acids, sugars and long-chain fatty acids by extracellular microbial enzymes. The hydrolysis of these complex organic polymers has been identified as the key to success of the anaerobic digestion process. It has been shown that the improvement in the hydrolysis step enhanced biogas production of anaerobic digestion. Many techniques have been reported to improve the performance of anaerobic digestion system, in particular,

Table 1 Anaerobic digestion process details

Step	Process description
Hydrolysis	Complex organic polymers such as carbohydrates, lipids and proteins are converted into dissolved monomers such as amino acids, sugars and long-chain fatty acids
Acidogenesis	The monomers are converted into volatile fatty acids and alcohols by acidogenic microorganisms. Hydrogen and carbon dioxide are also formed in this step
Acetogenesis	Organic acids are converted to acetate, carbon dioxide and hydrogen by acetogenic microorganism
Methanogenesis	Methanogenic microorganisms convert acetate, hydrogen and carbon dioxide into methane and carbon dioxide

the hydrolysis step. Among them is the addition of enzymes to accelerate the hydrolysis reaction. Pretreatment of the substrate has also been reported. The main purpose of pretreatment is to make the organic substrate more accessible to microbial action. Pretreatment methods can be classed into three major categories, namely, chemical, physical and biological as described in the following section.

Therefore, in this chapter, the utilisation of enzymes in hydrolysis is highlighted. This review will focus on the hydrolysis and pretreatment of substrates in anaerobic digestion mediated by enzymes. In addition, the effect of enzyme treatment combined with other types of pretreatment is also presented.

2 Role of Enzymes in Anaerobic Digestion

Enzymes are complex organic molecules made of protein and present in every living cell. As catalyst in biochemical reactions, enzymes decrease activation energy, thus speeding up the rate of reaction [6]. In recent years, there has been an increasing interest in the use of enzyme for synthesis of fine and commodity chemicals, pharmaceutical and agrochemical intermediates, and in waste treatment.

In anaerobic digestion, bacteria degrade substrates through enzymes. Endoenzymes and exoenzymes are two types of enzyme produced in the cell and involved in substrate degradation. Schematic diagrams of endoenzymes and exoenzymes are shown in Fig. 2. Endoenzymes are able to degrade soluble substrate within the cell. Meanwhile, exoenzymes are transported extracellularly through the slime coating, where they break down the insoluble substrate attached to the slime. Once in contact with the substrate, exoenzymes proceed to solubilise particulate and colloidal substrates. Then, these substrates enter the cell and are subsequently degraded by endoenzymes [2].

Not all bacteria produce exoenzymes. According to Burgess and Pletschke [7], exoenzymes can originate from one of three key sources, namely, sewage influent, activated sludge via cell autolysis and enzymes that are actively secreted by cells. Moreover, enzymes are known as reaction- or substrate-specific catalysts. Each

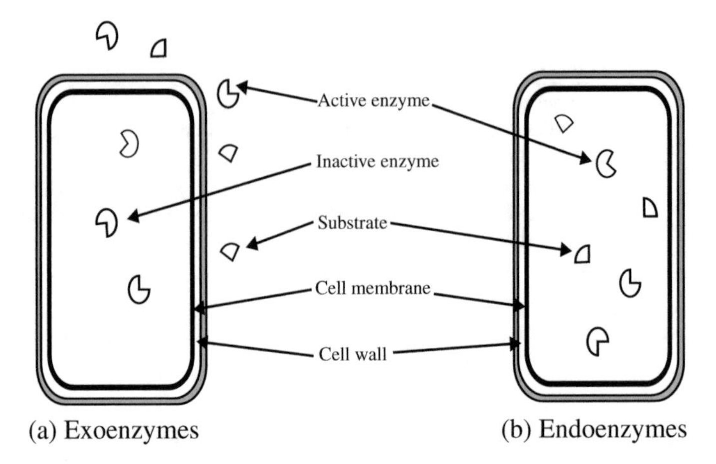

Fig. 2 Types of enzymes involved in substrate degradation [43]

enzyme degrades only a specific substrate or a group of substrates. In order to achieve a high degradation rate, it is important to provide proper types of exoenzymes and endoenzymes in a system. Therefore, for degradation of a wide variety of substrates, a large and diverse community of bacteria is needed [2]. Furthermore, bacteria and archaea involved in anaerobic digestion can only utilise simple compounds with low molecular weight. Additional enzymes are needed to break up larger molecules, for example, proteins, lipids and carbohydrates, which are among the major constituents in anaerobic digestion [8]. Hence, supplying additional enzymes into the anaerobic system is considered as a great alternative for stimulating waste degradation, thus, enhancing the performance of the anaerobic digestion system [9].

Besides that, enzymes are naturally occurring compounds that are biodegradable [10]. Their negligible contribution to biochemical oxygen demand (BOD) in the waste stream makes them harmless in anaerobic treatment processes and aquatic ecosystems [11]. In addition, by being versatile, enzymes can catalyse a variety of substrates under a wide range of environmental conditions. They can work in the presence of various types of microorganisms, recalcitrant and toxic substrates, as well as microbial metabolism inhibitors. Moreover, the enzymes' higher solubility and smaller size allow easier access to substrates compared to microbes [12]. These characteristics make it possible for enzymes to be applied in a wide range of applications, including anaerobic digestion.

3 Application of Enzymes During Hydrolysis Stage in Anaerobic Digestion

The first stage in anaerobic digestion is hydrolysis. During enzymatic hydrolysis, polymers are transformed into soluble monomers. This process is mediated by extracellular microbial enzymes known as hydrolases or lyases. Depending on the type of reaction catalysed, these hydrolases can be esterase, glycosidase or peptidase. For example, esterase and glycosidase are involved in hydrolysis of hemicellulose-based substrate, whereas peptidase is able to mediate the hydrolysis of substrate with high amount of protein [13]. In this stage, polysaccharides are transformed into simple sugars; cellulase enzyme-mediated hydrolysis of cellulose produces glucose, degradation of hemicellulose yields monosaccharide and amylase enzyme converts starch to glucose [2].

Many previous researches have shown that hydrolysis is a relatively slow process and generally, it limits the rate of the overall anaerobic digestion process [13]. Therefore, in order to improve the efficiency of biomass conversion in anaerobic digestion, increasing the hydrolysis rate is critical. Enzymatic hydrolysis is one of the available options to promote the performance of anaerobic digestion. This process can be operated under mild condition, has high substrate and reaction specificity, and generates no by-product [14, 15]. Some examples of enzyme application in hydrolysis of various substrates are presented in Table 2.

3.1 *Enzymatic Hydrolysis of Lignocellulosic Materials in Anaerobic Digestion*

The increase in interest is being shown in using enzymes in the hydrolysis of lignocellulose-rich substrates. Lignocellulosic materials are composed of cellulose and hemicellulose that are tightly bound to lignin. The addition of enzymes into this type of substrate could facilitate lignin degradation in the anaerobic digestion process. Lignin degradation increases hydrolysis rate by allowing microorganisms easy access to cellulose and hemicellulose in a lignocellulosic substrate. Thus, the enzymatic activity can be manipulated by supplying additional enzymes into the system [16, 17].

The ability of cellulase, hemicellulase and β-glucosidase enzymes to hydrolyse plant cell walls in anaerobic digestion systems has been studied by Romano et al. [17]. Anaerobic digestion was conducted in batch reactors operated at 50 °C, using wheatgrass as a model substrate. The results showed that when added directly to a single-stage digester, the enzymes had no significant effect on biogas production. On the other hand, enzyme addition during pretreatment of a two-stage digestion system effectively solubilised the wheatgrass leading to a higher biogas yield. Enzymatic pretreatment promotes hydrolysis of lignocellulose, breaking it down to lower molecular weight substances that are ready to be utilised by the bacteria. Nevertheless, at the end of the digestion period, there were no significant improvements in the biogas and methane yields and volatile solids reduction compared to the untreated reactors.

Fungal hydrolytic enzyme mixtures, with major components consisting of cellulase, hemicellulase, xylanase, pectinase, xylan esterase, pectinesterase, lipase, amylase glucosidase and protease, have been applied in different types of lignocellulose-rich feedstock. The results obtained depict a clear increase in

Table 2 Enzymatic hydrolysis of various substrates in anaerobic digestion system

Waste	Feedstock	Reactor type	Reactor configuration	Enzymes	References
Lignocellulosic waste	*Jose Tall* wheatgrass	Batch	Single and two stages	Novozyme N342, Celluclast C15L, and Novozyme N188 (cellulase, hemicellulase and β-glucosidase)	Romano et al. [17]
	Rye grain silage, maize silage, grass silage, feed residues and solid cattle manure	Batch	Single stage	Fungal hydrolytic enzyme mixture (cellulase, hemicellulase, xylanase, pectinase, xylan esterase, pectinesterase, lipase, amylase glucosidase and protease)	Quiñones et al. [10]
	Municipal solid waste	Batch	Single stage	Lignin peroxidase and manganese peroxidase	Hettiaratchi et al. [9]
	Hemp, flax, corn stover, *Miscanthus*, willow, ensilaged maize and wheat straw	Batch		Laccase from *Trametes versicolor*, peroxidase from *Bjerkandera adusta*	Schroyen et al. [18]
	Ensiled forage	Batch		Cellulolytic enzymes from methanogenic microbial community	Speda et al. [19]
Food waste	Scotch whisky distillery spent wash and pot ale	Batch	Two stages	Lyticase, α-amylase, cellulase, β-glucosidase, β-glucanase, lipase, protease and papain	Mallick et al. [20]
	Food waste from cafeteria	Upflow anaerobic sludge blanket		Carbohydrase from *Aspergillus* and *Aculeatus* Protease from *Aspergillus oryzae Lipase from Candida rugosa*	Moon and Song [21]
	Blended domestic food waste	Batch	Two stages	Fungal mash rich in hydrolytic	Uçkun Kiran et al. [22]
	Mixture of food waste and activated sludge	Batch	Two stages	Fungal mash rich in hydrolytic enzymes	Yin et al. [23]
	Chinese food waste	Batch	Two stages	Lipases from *Aspergillus*, *Candida* and Porcine pancreatic	Meng et al. [24]
	Wastewater from poultry industry	Batch	Single stage	Lipase from Porcine pancreatic	Dors et al. [26]

(continued)

Table 2 (continued)

Waste	Feedstock	Reactor type	Reactor configuration	Enzymes	References
	Milk fat	Batch		Lipases from *Geotrichum candidum* and *Candida rugosa*	Domingues et al. [27]
	Chicken feather waste	Batch	Two stages	Savinase 16 L (alkaline serine protease)	Forgács et al. [28]
	Coconut oil mill effluents	Batch	Two stages	Lipase from *Staphylococcus pasteuri*	Kanmani et al. [29]
Sewage sludge	Municipal wastewater	Batch	Single stage	α-amylase	Luo et al. [33]
	Primary sludge	Batch and continuous flow		Lipase, amylase and protease	Diak et al. [8]
	Wastewater treatment plant	Batch	Single stage	Lipase	Donoso-Bravo and Fdz-Polanco [11]

methane production for solid cattle manure, grass silage, feed residue and maize silage [10]. In another example, lignin peroxidase (LiP) and manganese peroxidase (MnP from *Phanerochaete chrysosporium*) successfully accelerated lignin-rich waste degradation and produced higher cumulative methane at the end of the digestion process [9]. Similarly, Schroyen et al. [18] found that enzymatic hydrolysis using laccase and peroxidase assisted lignin's matrix degradation, leading to a significantly higher release of total phenolic compounds from plant biomass.

Recently, the addition of exogenous enzymes into the digester to treat cellulosic materials has shown mixed success. The reason for this finding might be the origin of the enzymes, where the enzymes used originated from organisms. These enzymes were not evolutionarily adapted to the environment of anaerobic digesters. As a response to this problem, Speda et al. [19] investigated the potential of cellulolytic enzymes, which are produced and collected from a specific methanogenic microbial community. It was concluded that the enzymatic activity improved the yield and production rate of biogas.

3.2 Enzymatic Hydrolysis of Food Waste in Anaerobic Digestion

Food waste from households, restaurants and the food industry is rich in proteins, fats, carbohydrates and so forth, which are excellent substrates for anaerobic digestion. Wastewater from the food industry is difficult to treat through conventional biological treatment systems due to its chemical oxygen demand and high concentration of sugar, proteins, oil and grease [5].

In previous research, numerous yeast cell wall hydrolytic enzymes were studied on anaerobic digestion of Scotch whisky distillery spent wash and pot ale that have high oxygen demand. By conducting enzymatic pretreatment prior to anaerobic digestion of pot ale residues, the results showed chemical oxygen demand (COD) reductions of 87%, compared to 13% for digestion without enzymes [20].

Commercial enzymes such as carbohydrases, proteases and lipases have been used to improve hydrolysis of food waste. It was found that the mixture of these three enzymes appeared to be more efficient than that of using only a single commercial enzyme. This enzyme mixture resulted in high yield of methane and removal of soluble chemical oxygen demand [21].

Previous studies have shown that a fungal mash rich in hydrolytic enzymes successfully improved the hydrolysis rate of food waste for anaerobic digestion. For example, a fungal mash produced from cake waste, which has high amount of protease and glucoamylase, applied for enzymatic hydrolysis of mixed food waste, achieved a volatile solid removal of $80.4 \pm 3.5\%$. Moreover, the production rate and biomethane yield from food waste pretreated with fungal mash were found to be 3.5 and 2.3 times higher, respectively, than without pretreatment [22]. A similar type of enzyme source was used for pretreatment of food waste, activated sludge and their admixture to improve the efficiency of subsequent anaerobic digestion. The methane yield was found to be 2.5 for activated sludge mixed with food waste pretreated with fungal mash. This value is 1.6 times higher compared to activated sludge alone without and with pretreatment. Meanwhile, the total volatile solid was reduced by 54.3% [23]. Besides that, in a recent study, it was found that lipase addition could enhance biomethane production of high-lipid content food waste. Pretreatment using lipase shortened digestion time by 10–40 days while methane production rate was enhanced by 26.9–157.7% [24].

Furthermore, enzymatic pretreatment has been conducted on wastewater rich in lipids. It is known that the presence of lipid substrate in anaerobic digesters may lead to operational problems such as flotation of granular biomass conducting to washout and severe toxicity due to lipids affecting both methanogenic and acetogenic microorganisms [25]. The efficiency of applying porcine pancreatic lipase in simultaneous enzymatic hydrolysis and anaerobic biodegradation of lipid-rich wastewater from the poultry industry has been studied by Dors et al. [26]. The results showed that hydrolysis and anaerobic biodegradation could be carried out concurrently using a low enzyme concentration of 0.05 gL^{-1}. All samples pretreated with lipase displayed a positive effect on COD and colour removal as well as enhancement in methane formation. Another experiment on enzymatic treatment of lipid-rich substrate was conducted by Domingues et al. [27]. The authors discovered that lipase from *Candida rugosa* was suitable for use in hydrolysis of dairy waste as indicated by high biogas and specific methane productions.

A previous study showed that enzymatic pretreatment could greatly enhance the hydrolysis of chicken feather waste in anaerobic digestion [28]. Feather is composed of 90–92% protein, present in the form of β-keratin, which is very stable and has a relatively low biodegradability rate. Enzymatic hydrolysis mediated by serine protease was proven to improve the digestibility of feathers and enhance its biogas yield, where an increase of 122% in methane yield was achieved as compared to the untreated feathers.

Other than that, partially purified lipase from *Staphylococcus pasteuri* COM-4A immobilised on a Celite carrier has been applied for the enzymatic hydrolysis of unsterilised coconut oil mill effluent. Oil and grease and chemical oxygen demand were reduced successfully while the volatile fatty acid and long-chain fatty acid contents of the hydrolysed effluent increased significantly. Furthermore, anaerobic biodegradation of the prehydrolysed effluent showed significant improvement in biogas production and organic load removal [29].

The effect of *Carica papaya* latex and peel as an enzyme source on hydrogen yield and degradation efficiency of glucose, protein and lipid has been investigated. An increment in hydrogen yield based on protein and lipid degradation with substrate degradation efficiency of 51.3 ± 4.4% and 33.7 ± 2.6%, respectively, was achieved [30].

3.3 Enzymatic Hydrolysis of Sewage Sludge in Anaerobic Digestion

Digested sludge is recognised as a complex material constituting particulate material, microorganisms and extracellular polymeric substances that are excreted by these microorganisms. Problems associated with sludge include (1) presence of non-digestible material that can be inorganically bound to either carbon or slowly digestible organic compounds; (2) most of the organic compounds are located within cells produced during the activated sludge treatment process. The stable structure of the cell wall makes it resistant to biodegradation; and (3) presence of lignocellulosic materials in the sludge. These problems can be solved by increasing the hydrolysis rate of the sludge biomass into fermentable structures [5].

An enzymatic hydrolysis of sludge enhanced the hydrolysis yield under mild operating conditions, with low energy consumption and less by-product formation [31]. For the last three decades, various types of enzymes such as amylase, protease and endo-glycanases have been successfully applied in the hydrolysis of organic compounds presents in sludge [32]. Luo et al. [33] investigated the effects of enzyme dosage and temperature on waste activated sludge hydrolysis. It was found that an increase in amylase enzyme loading led to higher reduction in volatile suspended solid. The optimum enzyme loading was found to be 0.06 g amylase/g dry sludge. The results also showed that when the temperature increasing from 40 to 70 °C, the rate constant of α-amylase hydrolysis process increased from 0.106 to 0.215 h^{-1}, while the reaction activation energy for volatile suspended solid hydrolysis reduced from 62.72 to 20.19 kJ/mol. From these findings, it was concluded that α-amylase strongly enhanced the hydrolysis of waste activated sludge and higher temperature contributed to higher hydrolysis efficiency.

Various enzymes have been used to promote degradation of common wastewater and sludge constituents. The effect of lipase, cellulase, amylase and protease on anaerobic digestion of primary sludge under typical septic tank conditions has been studied [8]. Septic tanks can be considered as simple anaerobic digesters that achieve partial degradation of organic materials under ambient conditions. It was determined that enzymatic treatment did not increase hydrolysis and digestion rates in primary sludge. Moreover, no significant improvements were observed in removal of total solids, volatile solids, total suspended solids, total and soluble chemical oxygen demand and organic acids compared to control reactors.

Donoso-Bravo and Fdz-Polanco [11] assessed the effect of lipase addition and its dosage in anaerobic digestion of sewage sludge and grease traps. The addition of grease trap residue into the anaerobic digestion of sewage sludge has a negative effect on waste biodegradability. They discovered that enzyme addition notably increased methane production for all samples studied.

4 Enzymatic Pretreatment Combined with Other Physicochemical and Chemical Pretreatments

The main objective of pretreatment is to increase substrate solubility in order to accelerate the hydrolysis process and to decrease the amount of sludge for disposal. Different approaches have been applied to enhance the effectiveness of anaerobic digestion [12, 34]. Many chemical and physical treatments including thermal, pretreatment methods have been evaluated. One of them is the combination of pretreatment to increase biodegradability of waste. Some examples are shown in Table 3.

Enzymatic pretreatment can be combined with chemical treatment methods. Chemical and enzymatic sequential pretreatment of oat straw has been reported earlier [35]. The cellulose mediated hydrolysis of oat straw was conducted after acid and alkaline hydrolysis. It was found that solubilisation of 96.8% hemicellulose, 77.2% cellulose, and 42.2% lignin was achieved with the combination of mild acid and enzymatic pretreatment.

On the other hand, Rollini et al. [15] combined alkaline-enzymatic pretreatment of ensiled sorghum forage, a lignocellulosic substrate. Higher specific methane production rates were obtained compared to untreated samples. This combination showed a positive impact by solubilising up to 32 and 56% of cellulose and hemicellulose, respectively, compared to sole enzymatic hydrolysis. Furthermore, the enhancement of hydrolysis rate of the combination pretreatment may have resulted from physical redistribution or changes in composition of lignin and increase of accessible surface area and pore volume of the substrate after alkaline pretreatment.

Similar findings have been obtained by Michalska et al. [3] when alkaline and enzymatic pretreatment of energy crops was conducted. The authors concluded that the combined treatment in a two-step process is more efficient with regard to biogas production; it was about 30% higher and exceeded 60% volume of methane yield, in comparison to enzymatic hydrolysis alone. Alkaline treatment allows efficient delignification and partially solubilises hemicellulose and thus, increases accessibility of lignocelluloses for both enzymes and microorganisms. Moreover, the removal of hemicellulose and lignin was shown to enhance hydrolysis of lignocelluloses [36, 37].

Enzymatic treatment using lipase coupled with ultrasound irradiation was done in the pretreatment of synthetic dairy wastewater containing around 2000 mg/L fat [38]. By applying this technique, 78% hydrolysis yield was achieved with minimum exposure time due to the increase of mass transfer rate. A different finding was observed when this technique was applied for pretreatment of corn cob and vine trimming shoots, agricultural wastes with high lignocellulosic content [14]. The application of ultrasound and subsequent enzymatic hydrolysis improved biogas production from corn cob but not from vine trimming shoots due to the negative effects of ultrasound. Nonetheless, enzymatic pretreatment alone using Ultraflo enzyme successfully enhanced the transformation of both substrates into biogas.

Table 3 Pretreatment methods used in combination with enzymatic pretreatment

Pretreatment	Process	Substrate	Enzymes	Effect	References
Alkaline-acid-enzymatic pretreatments	Series	Oat straw	Celluclast 1.5L (cellulase)	Solubilization 96.8% of hemicellulose, 77.2% of cellulose and 42.2% of lignin	Gomez-Tovar et al. [35]
Alkaline-enzymatic pretreatments	Series	Ensiled sorghum forage	– Agazym BGL (cellulose, β-glucanase, hemicellulase and xylanase from *Aspergillus aculeatus*) – Ultra L (polygalacturonase and pectinase from *Aspergillus* strains) – Pulpzyme HC (endo-xylanases from *Bacillus* strains) – Primafast 200 (endo-1-4-β-glucanases)	– Solubilization 32% cellulose and 56% hemicelluloses – 37% increase in methane production	Rollini et al. [15]
Alkaline-enzymatic pretreatments	Series	Energy crops (*Miscanthus giganteus* and *Sida hermaphrodita*)	Celluclast 1.5L(cellulase) and Novozyme 188 (cellobiase)	60% increase in methane yield	Michalska et al. [3]
Ultrasound-enzymatic pretreatment	In situ	Dairy wastewater	Lipase from *Candida rugosa*	– Reduces the reaction time from 24 h to 40 min – 78% hydrolysis rate	Adulkar and Rathod [38]
Ultrasound-enzymatic pretreatment	Series	Corn cob and vine trimming shoots	Ultraflo® L(endo-1,3(4)-β-glucanase, and collateral xylanase, cellobiase, cellulase and feruloyl esterase)	– Improved the biogas production from corn cob – Reduced biogas production from vine trimming shoots	Pérez-Rodríguez et al. [14]
Grinding-thermal-enzymatic pretreatment	Series	Sugar beet pulp	Celustar XL (endoglucanase, xylanase) and Agropect pomace (pectinase)	Highest cumulative biogas productivity of 898.7 mL/gVS with combination of all three method	Ziemiński and Kowalska-Wentel [39]
Alkali extrusion-enzymatic pretreatment	Series	Corn cob	Ultraflo® L(endo-1,3(4)-β-glucanase, and collateral xylanase, cellobiase, cellulase and feruloyl esterase) and *Aspergillus* enzymes extract	Increase of the methane volume produced by 22.3%	Pérez-Rodríguez et al. [42]

The combination of mechanical and thermal treatments prior to enzymatic treatment could reduce cost. Ziemiński and Kowalska-Wentel [39] investigated the effect of different sugar beet pulp pretreatments on biogas yield in anaerobic digestion. Celustar XL (endoglucanase, xylanase) and Agropect pomace (pectinase) were used in the enzymatic pretreatment. It was observed that the enzymatic hydrolysates of the ground and thermal-pressure pretreated substrates achieved the highest cumulative biogas productivity of 898.7 mL/g volatile solid. It is stated in the literature that grinding of lignocellulosic materials affects the structure of the cellulose. After grinding, it is more susceptible to enzymatic depolymerisation due to the reduced in the crystallinity of cellulose [40].

Among various pretreatment processes, extrusion is simple, flexible and adaptable. Extrusion is a process in which uniformly moistened biomass material is passed through an extruder barrel applying pressure with a screw. This method has been reported as an effective physical method for biomass size reduction that enhances anaerobic digestion of lignocellulosic materials [41]. Pérez-Rodríguez et al. [42] have investigated the effect of this pretreatment alone and in combination with alkaline and/or enzymatic hydrolysis to improve production of methane via anaerobic digestion of corn cob. Among all the pretreatments studied, sequential alkaline extrusion and enzymatic hydrolysis pretreatment achieved the highest methane production rate.

5 Conclusion

This review clearly indicates that utilisation of enzyme is an alternative to overcome limitations associated with anaerobic digestion. The effect of enzymatic hydrolysis in digester and pretreatments is, however, very dependent on biomass composition and operating conditions. Future research on development and optimisation of enzymatic hydrolysis is needed to achieve maximum benefit from this technology. In addition, to reduce the cost of treatment using enzymes, more investigation on enzyme preparation, stability and activity should be conducted.

References

1. Pilli S, More TT, Yan S, Tyagi RD, Surampalli RY, Zhang TC (2016) Anaerobic digestion or co-digestion for sustainable solid waste treatment/management. In: Sustainable solid waste management, pp 187–232. https://doi.org/10.1061/9780784414101.ch08
2. Merlin Christy P, Gopinath LR, Divya D (2014) A review on anaerobic decomposition and enhancement of biogas production through enzymes and microorganisms. Renew Sustain Energy Rev 34:167–173. https://doi.org/10.1016/j.rser.2014.03.010
3. Michalska K, Bizukojć M, Ledakowicz S (2015) Pretreatment of energy crops with sodium hydroxide and cellulolytic enzymes to increase biogas production. Biomass Bioenergy 80:213–221. https://doi.org/10.1016/j.biombioe.2015.05.022

4. Madsen M, Holm-Nielsen JB, Esbensen KH (2011) Monitoring of anaerobic digestion processes: a review perspective. Renew Sustain Energy Rev 15(6):3141–3155. https://doi.org/10.1016/j.rser.2011.04.026
5. Parawira W (2012) Enzyme research and applications in biotechnological intensification of biogas production. Crit Rev Biotechnol 32(2):172–186. https://doi.org/10.3109/07388551.2011.595384
6. Quiñones TS, Plöchl M, Päzolt K, Budde J, Kausmann R, Nettmann E, Heiermann M (2012) Hydrolytic enzymes enhancing anaerobic digestion. In: Biogas production: pretreatment methods in anaerobic digestion, pp 157–198. https://doi.org/10.1002/9781118404089.ch6
7. Burgess JE, Pletschke BI (2008) Hydrolytic enzymes in sewage sludge treatment: a mini-review. Water SA 34(3):343–349
8. Diak J, Örmeci B, Kennedy KJ (2012) Effect of enzymes on anaerobic digestion of primary sludge and septic tank performance. Bioprocess Biosyst Eng 35(9):1577–1589. https://doi.org/10.1007/s00449-012-0748-7
9. Hettiaratchi JPA, Jayasinghe PA, Bartholameuz EM, Kumar S (2014) Waste degradation and gas production with enzymatic enhancement in anaerobic and aerobic landfill bioreactors. Bioresour Technol 159:433–436. https://doi.org/10.1016/j.biortech.2014.03.026
10. Quiñones TS, Plöchl M, Budde J, Heiermann M (2012) Results of batch anaerobic digestion test - effect of enzyme addition. Agricult Eng Int: CIGR J 14(1):38–50
11. Donoso-Bravo A, Fdz-Polanco M (2013) Anaerobic co-digestion of sewage sludge and grease trap: assessment of enzyme addition. Process Biochem 48(5–6):936–940. https://doi.org/10.1016/j.procbio.2013.04.005
12. Romero-Güiza MS, Vila J, Mata-Alvarez J, Chimenos JM, Astals S (2016) The role of additives on anaerobic digestion: a review. Renew Sustain Energy Rev 58:1486–1499. https://doi.org/10.1016/j.rser.2015.12.094
13. Kondusamy D, Kalamdhad AS (2014) Pre-treatment and anaerobic digestion of food waste for high rate methane production – a review. J Environ Chem Eng 2(3):1821–1830. https://doi.org/10.1016/j.jece.2014.07.024
14. Pérez-Rodríguez N, García-Bernet D, Domínguez JM (2016) Effects of enzymatic hydrolysis and ultrasounds pretreatments on corn cob and vine trimming shoots for biogas production. Bioresour Technol 221:130–138. https://doi.org/10.1016/j.biortech.2016.09.013
15. Rollini M, Sambusiti C, Musatti A, Ficara E, Retinò I, Malpei F (2014) Comparative performance of enzymatic and combined alkaline-enzymatic pretreatments on methane production from ensiled sorghum forage. Bioprocess Biosyst Eng 37(12):2587–2595. https://doi.org/10.1007/s00449-014-1235-0
16. Jayasinghe PA, Hettiaratchi JPA, Mehrotra AK, Steele MA (2013) Enhancing gas production in landfill bioreactors: flow-through column study on leachate augmentation with enzyme. J Hazard Toxic Radioact Waste 17(4):253–258. https://doi.org/10.1061/(ASCE)HZ.2153-5515.0000166
17. Romano RT, Zhang R, Teter S, McGarvey JA (2009) The effect of enzyme addition on anaerobic digestion of Jose Tall Wheat Grass. Bioresour Technol 100(20):4564–4571. https://doi.org/10.1016/j.biortech.2008.12.065
18. Schroyen M, Vervaeren H, Vandepitte H, Van Hulle SWH, Raes K (2015) Effect of enzymatic pretreatment of various lignocellulosic substrates on production of phenolic compounds and biomethane potential. Bioresour Technol 192:696–702. https://doi.org/10.1016/j.biortech.2015.06.051
19. Speda J, Johansson MA, Odnell A, Karlsson M (2017) Enhanced biomethane production rate and yield from lignocellulosic ensiled forage ley by in situ anaerobic digestion treatment with endogenous cellulolytic enzymes. Biotechnol Biofuels 10(1). https://doi.org/10.1186/s13068-017-0814-0
20. Mallick P, Akunna JC, Walker GM (2010) Anaerobic digestion of distillery spent wash: influence of enzymatic pre-treatment of intact yeast cells. Bioresour Technol 101(6):1681–1685. https://doi.org/10.1016/j.biortech.2009.09.089

21. Moon HC, Song IS (2011) Enzymatic hydrolysis of foodwaste and methane production using UASB bioreactor. Int J Green Energy 8(3):361–371. https://doi.org/10.1080/15435075.2011.557845
22. Uçkun Kiran E, Trzcinski AP, Liu Y (2015) Enhancing the hydrolysis and methane production potential of mixed food waste by an effective enzymatic pretreatment. Bioresour Technol 183:47–52. https://doi.org/10.1016/j.biortech.2015.02.033
23. Yin Y, Liu YJ, Meng SJ, Kiran EU, Liu Y (2016) Enzymatic pretreatment of activated sludge, food waste and their mixture for enhanced bioenergy recovery and waste volume reduction via anaerobic digestion. Appl Energy 179:1131–1137. https://doi.org/10.1016/j.apenergy.2016.07.083
24. Meng Y, Luan F, Yuan H, Chen X, Li X (2017) Enhancing anaerobic digestion performance of crude lipid in food waste by enzymatic pretreatment. Bioresour Technol 224:48–55. https://doi.org/10.1016/j.biortech.2016.10.052
25. Angelidaki I, Sanders W (2004) Assessment of the anaerobic biodegradability of macropollutants. Rev Environ Sci Biotechnol 3(2):117–129. https://doi.org/10.1007/s11157-004-2502-3
26. Dors G, Mendes AA, Pereira EB, de Castro HF, Furigo A (2013) Simultaneous enzymatic hydrolysis and anaerobic biodegradation of lipid-rich wastewater from poultry industry. Appl Water Sci 3(1):343–349. https://doi.org/10.1007/s13201-012-0075-9
27. Domingues RF, Sanches T, Silva GS, Bueno BE, Ribeiro R, Kamimura ES, Franzolin Neto R, Tommaso G (2015) Effect of enzymatic pretreatment on the anaerobic digestion of milk fat for biogas production. Food Res Int 73:26–30. https://doi.org/10.1016/j.foodres.2015.03.027
28. Forgács G, Lundin M, Taherzadeh MJ, Horváth IS (2013) Pretreatment of chicken feather waste for improved biogas production. Appl Biochem Biotechnol 169(7):2016–2028. https://doi.org/10.1007/s12010-013-0116-3
29. Kanmani P, Kumaresan K, Aravind J (2015) Pretreatment of coconut mill effluent using celite-immobilized hydrolytic enzyme preparation from *Staphylococcus pasteuri* and its impact on anaerobic digestion. Biotechnol Prog 31(5):1249–1258. https://doi.org/10.1002/btpr.2120
30. Elsamadony M, Tawfik A, Danial A, Suzuki M (2015) Use of *Carica papaya* enzymes for enhancement of H_2 production and degradation of glucose, protein, and lipids. Energy Proc 975–980. https://doi.org/10.1016/j.egypro.2015.07.308
31. Chen YT, Wang FS (2011) Determination of kinetic parameters for enzymatic cellulose hydrolysis using hybrid differential evolution. Int J Chem React Eng 9
32. Yang Q, Luo K, Xm Li, Db Wang, Zheng W, Gm Zeng, Jj Liu (2010) Enhanced efficiency of biological excess sludge hydrolysis under anaerobic digestion by additional enzymes. Bioresour Technol 101(9):2924–2930. https://doi.org/10.1016/j.biortech.2009.11.012
33. Luo K, Yang Q, Li XM, Yang GJ, Liu Y, Wang DB, Zheng W, Zeng GM (2012) Hydrolysis kinetics in anaerobic digestion of waste activated sludge enhanced by α-amylase. Biochem Eng J 62:17–21. https://doi.org/10.1016/j.bej.2011.12.009
34. Jha AK, Li J, Nies L, Zhang L (2011) Research advances in dry anaerobic digestion process of solid organic wastes. Afr J Biotechnol 10(65):14242–14253
35. Gomez-Tovar F, Celis LB, Razo-Flores E, Alatriste-Mondragón F (2012) Chemical and enzymatic sequential pretreatment of oat straw for methane production. Bioresour Technol 116:372–378. https://doi.org/10.1016/j.biortech.2012.03.109
36. Safari A, Karimi K, Shafiei M (2017) Dilute alkali pretreatment of softwood pine: a biorefinery approach. Bioresour Technol 234:67–76. https://doi.org/10.1016/j.biortech.2017.03.030
37. Hendriks ATWM, Zeeman G (2009) Pretreatments to enhance the digestibility of lignocellulosic biomass. Bioresour Technol 100(1):10–18. https://doi.org/10.1016/j.biortech.2008.05.027
38. Adulkar TV, Rathod VK (2014) Ultrasound assisted enzymatic pre-treatment of high fat content dairy wastewater. Ultrason Sonochem 21(3):1083–1089. https://doi.org/10.1016/j.ultsonch.2013.11.017

39. Ziemiński K, Kowalska-Wentel M (2017) Effect of different sugar beet pulp pretreatments on biogas production efficiency. Appl Biochem Biotechnol 181(3):1211–1227. https://doi.org/10.1007/s12010-016-2279-1
40. Krishania M, Kumar V, Vijay VK, Malik A (2013) Analysis of different techniques used for improvement of biomethanation process: a review. Fuel 106:1–9. https://doi.org/10.1016/j.fuel.2012.12.007
41. Ravindran R, Jaiswal AK (2016) A comprehensive review on pre-treatment strategy for lignocellulosic food industry waste: challenges and opportunities. Bioresour Technol 199:92–102. https://doi.org/10.1016/j.biortech.2015.07.106
42. Pérez-Rodríguez N, García-Bernet D, Domínguez JM (2017) Extrusion and enzymatic hydrolysis as pretreatments on corn cob for biogas production. Renew Energy 107:597–603. https://doi.org/10.1016/j.renene.2017.02.030
43. Talaro KP, Talaro A (2002) Foundations in microbiology: basic principles. McGraw-Hill

Process Simulation of Anaerobic Digestion Process for Municipal Solid Waste Treatment

Noorlisa Harun, Wan Hanisah W. Ibrahim, Muhamad Faez Lukman, Muhammad Hafizuddin Mat Yusoff, Nur Fathin Shamirah Daud and Norazwina Zainol

Abstract A simulation of the anaerobic digestion process for municipal solid waste (MSW) treatment has been carried out using Aspen Plus software. Anaerobic digestion uses enzymes to solubilise particulate organic compounds so that they can be easily separated from inert waste such as plastic, metals and textiles. The complex substrates such as proteins, carbohydrates and fats are hydrolyzed into their respective monomers, such as amino acids, glucose and fatty acids. The hydrolyzed monomers then are converted into different volatile fatty acids (VFAs); later the VFAs are converted into carbon dioxide, acetic acid and hydrogen. A model of the anaerobic digestion process is represented by RSTOIC and RCSTR reactors in Aspen Plus. The hydrolysis reactions occur in RSTOIC reactor; meanwhile, amino acid degradation, acidogenic and acetogenic reactions are implemented in RCSTR reactor. The amount of dry matter content in bioliquid was 20 wt% which mainly consists of VFA. Sensitivity analysis has been performed in order to study the effect of residence time for the production of organic liquid fraction (bioliquid). The amount of bioliquid produced was increased as residence time was increased.

Keywords Anaerobic digestion · Process simulation · Bioliquid
Aspen Plus · Municipal solid waste · Food waste

1 Introduction

Municipal solid waste (MSW), commonly known as trash, garbage or rubbish, is discarded from residential, commercial and institutional areas, and consists of everyday items. MSW is being generated at a rate that exceeds the ability of the

N. Harun (✉) · W. H. W. Ibrahim · M. F. Lukman · M. H. M. Yusoff
N. F. S. Daud · N. Zainol
Faculty of Chemical and Natural Resources Engineering, Universiti Malaysia Pahang,
Lebuhraya Tun Razak, 26300 Gambang, Kuantan, Pahang, Malaysia
e-mail: noorlisa@ump.edu.my

N. Horan et al. (eds.), *Anaerobic Digestion Processes*,
Green Energy and Technology, https://doi.org/10.1007/978-981-10-8129-3_5

natural environment to naturalize it and municipal authorities to manage it, as the global population increases dramatically and with changing consumption patterns, economic development, rapid urbanization and industrialization. The situation is more severe in developing countries such as Malaysia where the rapid growth of the economy and population have caused MSW to proliferate by 28% in a period of a decade [1]. According to Malaysia Second National Communication to the UNFCCC [2], it is predicted to increase further by 30% in 2020 and 39% in 2030 compared to the baseline year of 2007. About 93.5% of municipal solid waste in Malaysia goes to landfills or open dumpsites without gas recovery, meanwhile, only 5.5% of MSW is recycled and 1.0% is composted [3]. The over-reliance on landfilling and inappropriate waste disposal has been continuously pressing the environmental, health and safety issues for the Malaysian citizens. It is also amplifying the share of total global anthropogenic greenhouse gas (GHG) emission, which is caused by the production of methane gas (CH_4) through the anaerobic decomposition of solid waste in landfills [4].

The Government of Malaysia is seeking practical solutions to improve the current waste management situation, including the sanitation and closure of illegal landfills, waste incineration with energy recovery, upgrading landfills with CH_4 recovery, composting of organic waste and recycling and waste minimisation. Among all the proposals, waste to energy (WTE) is a promising alternative to surmount the problem of waste generation and a potential renewable energy (RE) source for Malaysia [4]. WTE refers to the recovery or generation of the energy from waste materials into useable heat, electricity or fuel from the primary treatment of waste. MSW has the potential to bring new financial advantages and sources of fuel for future energy needs [5]. WTE has been practiced in Malaysia in recent decades and is implemented for biomass from agricultural waste and forestry residues (i.e. palm oil biomass, paddy straw and logging residues). However, WTE from MSW is still underutilized in Malaysia. Feasibility analyses of WTE from MSW in Malaysia have been conducted by local researchers over the past decade [6]. Besides MSW, WTE technologies are also applicable to different waste categories such as solid, liquid (e.g. domestic sewage), and gaseous (e.g. refinery flue gas). WTE approaches can be categorized into three types: thermal treatment, biological treatment and landfill. Biological treatment of WTE included anaerobic digestion process with the production of biogas [5].

The anaerobic digestion process occurs in environments depleted of oxygen and involves the breakdown of organic matter into biogas and other traces gases, as well as a residual effluent or digestate. There are four steps in the anaerobic digestion process: hydrolysis, acidogenesis, acetogenesis and methanogenesis. As a result, four steps of reactions (in the anaerobic digestion) can be differentiated due to the different kinds of microbial populations and the specific optimal parameters needed for each step. Although the anaerobic digestion process can potentially offer many benefits, it is difficult to implement because it is a complex system of biochemical and physical processes. Therefore, this requires a detailed understanding of design, control, operation and maintenance to ensure high process efficiency. For instance, the system can become unstable or even fail due to overload, accumulation of intermediate products, unsteady pH, lack of nutrients or key trace elements [7].

The main composition of municipal solid waste is food waste. Food waste (FW) is an organic waste discharged from various sources including food processing plants, domestic and commercial kitchens, cafeterias and restaurants. Food waste is a typical form of organic matter with a high potential for energy production through anaerobic degradation. Food waste is mainly composed of carbohydrate polymers (starch, cellulose and hemicelluloses), lignin, proteins, lipids, organic acids and a remaining, smaller inorganic part. Because of the benefits in terms of energy saving, waste management and environmental aspects, biogas production from food waste together with other renewable organic sources, i.e. agriculture waste has been suggested as a means of meeting one-third of renewable energy demand in transport by 2020 in the EU [7].

Food waste, like many other waste types, presents its own problems for the anaerobic digestion process. For example, the high protein content in food waste leads to high ammonia concentrations, which can be inhibitory to microorganism involved in the process and propionic acid can accumulate in the digester due to inadequate removal of formic acid or hydrogen as an intermediate. With certain types of feedstock, foaming problems can occur, which in severe cases may even cause digester failure [8–11].

This study aims to model the anaerobic digestion process to transform food waste into bioliquid. Although it can never replace experimental works, modelling is a useful aid in research as it helps to predict the production of bioliquid as well as to optimize the overall performance of anaerobic systems with respect to mass and energy balances, or for design and control purpose [12]. The model expected can assist in predicting the process output at various waste compositions. Generally, the main reasons for using a model would be considered as understanding the system's behaviour and the interaction of its components, quantitatively expressing or verifying hypothesis, and predicting the behaviour of the system in the future [12].

Aspen Plus software is used as a tool to model the anaerobic digestion process. It has rigorous methods for estimating the properties of components and meticulous thermodynamic calculations. Anaerobic Digestion Model No. 1 (ADM1), which is widely used and considered as the most complete model, with high accuracy in terms of data, reactions and kinetics calculations was also employed in this study. The ADM1 structured model provides multiple steps describing biochemical and physicochemical processes. The biochemical reactions include the hydrolysis of carbohydrates, proteins and lipids to sugars, amino acids and long-chain fatty acids (LCFA), respectively; acidogenesis from sugars and amino acids to volatile fatty acids (VFAs) and hydrogen; acetogenesis of LCFA and VFAs to acetate; and separate methanogenesis steps from acetate and hydrogen/CO_2 [13]. It has been accepted as the standard model for the anaerobic digestion of sludge and solid waste in terms of process design and dynamic simulation [12].

2 Methodology

In implementing the Aspen Plus simulation on the enzymatic reactor, a few steps need to be taken. Using this software, the operating conditions of digestion, which depends on substrate degradation, bioliquid production kinetics and yield materials, could be optimized. The model implementations consisted of property method selection, component list, reaction list and flowsheet synthesis.

2.1 Property Methods

Proper property method selection is important because inadequate selection can undermine the accuracy and prevent the simulation to be executed. The choice can strongly affect the prediction of the simulation. There are several property methods available in Aspen Plus software:

- GRAYSON: Recommended when hydrogen reaction is included.
- Peng Robinson: Useful for gas processing coupled with binary parameters.
- NRTL (recommended): Activity coefficients are taken into account.

The property method suitable for this anaerobic digestion process is the NRTL method. It is because this process included ionic molecules from the reaction in digestion process, and also the reaction involves with phase changes that occur throughout the simulation.

2.2 Component List

The component list consists of compounds that have to be filled in Aspen Plus. These components are specified as follows [14, 15]:

- Carbohydrate is represented as starch, cellulose, hemicellulose and xylose. Starch is represented as cellulose as it is not present in the Aspen Plus databank. Hemicellulose is represented by glutaric acid. Xylose and cellulose are represented by their own components.
- Fat components are represented by triolein, tripalmitin, palmito-olein and palmito-linolein. Some components of the fats do not exist in the Aspen Plus databanks. Palmito-olein and palmito-linolein are represented by Sn-1-Palmito-2-Olein and Sn-1-Palmito-2-Linolein, respectively.
- Protein is represented as soluble and insoluble Protein. The soluble protein is represented as protein while the insoluble protein is represented by keratin. Both components are represented as pseudocomponents in Aspen Plus.
- VFAs are represented by acetic acid, propionic acid, butyric acid, valeric acid, palmitic acid and linoleic acid. Palmitic acid is represented in the simulation as

1-Hexadecanol. The butyric and valeric acid is represented as isobutyric acid and isovaleric acid, respectively.
- Glucose is represented as dextrose in the simulation.
- Ethanol, water and ammonia exist as input components.
- All of the 20 amino acids, i.e. asparagine, glutamine, arginine, histidine, lysine, tyrosine, tryptophan, phenylalanine, cysteine, methionine, threonine, serine, leucine, isoleucine, valine, glutamic acid, aspartic acid, glycine, alanine and proline are also incorporated in the simulation.
- Other components which exist due to the reactions are benzene, phenol, carbonic acid, furfural, hydrogen sulphide, methyl mercaptan, formamide, indole, methane, hydrogen, carbon dioxide and ethyl cyanoacetate.
- Some components exist due to the acid–base reactions. These components are H^+, OH^-, NH_4^+, carbonic acid, HCO_3^-, CO_3^{-2} and HS^-.

Some of the components such as amino acids do not have complete property data in the Aspen Plus databank. Thus, property data for these components are obtained from the existing component in Aspen Plus data bank that has similar chemical and physical properties. For example, data for lysine was obtained from arginine due to the similarity in chemical and physical structures. Thermodynamics data of some chemical compounds, which is unavailable in the data bank, could be filled with the data of their similar chemical compounds [14].

2.3 List of Reactions

The list of hydrolysis reactions is stated in Table 1 while for amino acids degradation, acidogenesis and acetogenesis are listed in Table 2. The breakdowns of the larger molecules occur at the hydrolysis reaction. While the production of acetate and most of the VFA is at the other three reactions.

2.4 Flowsheet of Anaerobic Digestion Process

The next step was to synthesize the process flowsheet. The model used in this work is obtained from [15] with some modifications. The methanogenesis step is excluded in the current model to consider bioliquid production and the input data for the simulation was obtained from NADNO (2012). The original model of anaerobic digestion to produce biogas from organic waste is available at the Swedish database http://hdl.handle.net/2320/12358 [15].

In this simulation, two types of reactor are used, namely, the RSTOIC and RCSTR reactor. The reactors represent the digester where all the chemical and biochemical reactions occur.

Table 1 List of hydrolysis reactions [15]

No.	Compound	Hydrolysis reaction	Extent of reaction
1.	Starch	$(C_6H_{12}O_6)_n + H_2O \rightarrow n\ C_6H_{12}O_6$	0.6 ± 0.2
2.	Cellulose	$(C_6H_{12}O_6)_n + H_2O \rightarrow n\ C_6H_{12}O_6$	0.4 ± 0.1
3.	Hemicellulose	$C_5H_8O_4\ H_2O \rightarrow 2.5\ C_2H_4O_2$	0.5 ± 0.2
4.	Hemicellulose	$C_5H_8O_4\ H_2O \rightarrow C_5H_{10}O_5$	0.6 ± 0.0
5.	Xylose	$C_5H_{10}O_5 \rightarrow C_5H_4O_2 + 3\ H_2O$	0.6 ± 0.0
6.	Cellulose	$C_6H_{12}O_6 + H_2O \rightarrow 2\ C_2H_6O + 2\ CO_2$	0.4 ± 0.1
7.	Ethanol	$2\ C_2H_6O + CO_2 \rightarrow 2\ C_2H_4O_2 + CH_4$	0.6 ± 0.1
8.	Soluble protein	$C_{13}H_{25}O_7N_3S + 6\ H_2O \rightarrow 6.5\ CO_2 + 6.5\ CH_4 + 3\ H_3N + H_2S$	0.5 ± 0.2
9.	Insoluble protein (LP)	$I.P + 0.3337\ H_2O \rightarrow 0.045\ C_6H_{14}N_4O_2 + 0.048\ C_4H_7NO_4 + 0.047\ C_4H_9NO_3 + 0.172\ C_3H_7NO_3 + 0.074\ C_5H_9NO_4 + 0.111\ C_5H_9NO_2 + 0.25\ C_2H_5NO_2 + 0.047\ C_3H_7NO_2 + 0.067\ C_3H_6NO_2S + 0.074\ C_5H_{11}NO_2 + 0.07\ C_6H_{13}NO_2 + 0.046\ C_6H_{13}NO_2 + 0.036\ C_9H_{11}NO_2$	0.6 ± 0.1
10.	Triolein	$C_7H_{104}O_6 + 3H_2O \rightarrow C_3H_8O_3 + 3\ C_{18}H_{34}O_2$	0.5 ± 0.2
11.	Tripalmitin	$C_{51}H_{98}O_6 + 8.436\ H_2O \rightarrow 4\ C_3H_8O_3 + 2.43\ C_{16}H_{34}O$	0.5 ± 0.3
12.	Palmito-olein	$C_{37}H_{70}O_5 + 4.1\ H_2O \rightarrow 2.1\ C_3H_8O_3 + 0.9\ C_{16}H_{34}O + 0.9\ C_{18}H_{34}O_2$	0.6 ± 0.2
13.	Palmito-linolein	$C_{37}H_{68}O_5 + 4.3\ H_2O \rightarrow 2.2\ C_3H_8O_3 + 0.9\ C_{16}H_{34}O + 0.9\ C_{18}H_{32}O_2$	0.6 ± 0.2

- The first reactor used in the simulation is the RSTOIC reactor to represent hydrolysis reactions, which involved the breakdown of the larger components into simple monomers. The hydrolysis reactions are modelled using the extent of reaction method with the fractional conversion of reactants into products on a scale of 0.0–1.0. The stoichiometric reaction correlation for the hydrolysis reactions as listed in Table 1 is specified in Aspen Plus. The temperature and pressure of this reactor are specified at 328.15 K and 102,300 N/m^2, respectively.
- Next, the process is continued with the other type of reactor which is the RCSTR reactor. The function of RCSTR reactor is for the reaction of the product released from the previous reactor (RSTOIC) to take place and produced the desired final product which is the bioliquid. The reactions that occur in the RCSTR reactor are acetogenesis and acidogenesis. The reaction is assumed well mixed and the residence time that the input spends in the reactor is specified at 12 h during the simulation. The valid phase is set as liquid only. The temperature and pressure of this reactor are set at 328.15 K and 101,325 N/m^2, respectively. The volume of this reactor is 275 m^3 [16].

Table 2 List of amino acid, acidogenic and acetogenic reactions [15]

No.	Compound	Chemical reactions	Kinetic constant
Amino acid degradation reactions			
1.	Glycine	$C_2H_5NO_2 + H_2 \rightarrow C_2H_4O_2 + H_3N$	1.28×10^{-02}
2.	Threonine	$C_4H_9NO_3 + H_2 \rightarrow C_2H_4O_2 + 0.5\ C_4H_8O_2 + H_3N$	1.28×10^{-02}
3.	Histidine	$C_6H_8N_3O_2 + 4\ H_2O + 0.5\ H_2 \rightarrow CH_3NO + C_2H_4O_2 + 0.5\ C_4H_8O_2 + 2\ H_3N + CO_2$	1.28×10^{-02}
4.	Arginine	$C_6H_{14}N_2O + 3\ H_2O + H_2 \rightarrow 0.5\ C_2H_4O_2 + 0.5\ C_3H_6O_2 + 0.5\ C_5H_{10}O_2 + 4\ H_3N + CO_2$	1.28×10^{-02}
5.	Proline	$C_5H_9NO_2 + H_2O + H_2 \rightarrow 0.5\ C_2H_4O + 0.5\ C_3H_6O_2 + 0.5\ C_5H_{10}O_2 + H_3N$	1.28×10^{-02}
6.	Methionine	$C_5H_{11}NO_2S + 2\ H_2O \rightarrow C_3H_6O_2 + CO_2 + H_3N + H_2 + CH_4S$	1.28×10^{-02}
7.	Serine	$C_3H_7NO_3 + H_2O \rightarrow C_2H_4O_2 + H_3N + CO_2 + H_2$	1.28×10^{-02}
8.	Threonine	$C_4H_9NO_3 + H_2O \rightarrow C_3H_6O_2 + H_3N + H_2 + CO_2$	1.28×10^{-02}
9.	Aspartic acid	$C_4H_7NO_4 + 2\ H_2O \rightarrow C_2H_4O_2 + H_3N + 2\ CO_2 + 2\ H_2$	1.28×10^{-02}
10.	Glutamic acid	$C_5H_9NO_4 + H_2O \rightarrow C_2H_4O_2 + 0.4\ C_4H_8O_2 + H_3N + CO_2$	1.28×10^{-02}
11.	Glutamic acid	$C_5H_9NO_4 + 2\ H_2O \rightarrow 2\ C_2H_4O_2 + H_3N + CO_2 + H_2$	1.28×10^{-02}
12.	Histidine	$C_6H_8N_3O_2 + 5\ H_2O \rightarrow CH_3NO + 2\ C_2H_4O_2 + 2\ H_3N + CO_2 + 0.5\ H_2$	1.28×10^{-02}
13.	Arginine	$C_6H_{14}N_3O_2 + 6\ H_2O \rightarrow 2\ C_2H_4O_2 + 4\ H_3N + 2\ CO_2 + 3\ H_2$	1.28×10^{-02}
14.	Lysine	$C_6H_{14}N_2O_2 + 2\ H_2O \rightarrow C_2H_4O_2 + C_4H_8O_2 + 2\ H_3N$	1.28×10^{-02}
15.	Leucine	$C_6H_{13}NO_2 + 2\ H_2O \rightarrow C_5H_{10}O_2 + H_3N + CO_2 + 2\ H_2$	1.28×10^{-02}
16.	Isoleucine	$C_6H_{13}NO_2 + 2\ H_2O \rightarrow C_5H_{10}O_2 + H_3N + CO_2 + 2\ H_2$	1.28×10^{-02}
17.	Valine	$C_5H_{11}NO_2 + 2\ H_2O \rightarrow C_4H_8O_2 + H_3N + CO_2 + 2\ H_2$	1.28×10^{-02}
18.	Phenylalanine	$C_9H_{11}NO_2 + 2\ H_2O \rightarrow C_6H_6 + C_2H_4O_2 + H_3N + CO_2 + H_2$	1.28×10^{-02}
19.	Tyrosine	$C_9H_{11}NO_3 + 2\ H_2O \rightarrow C_6H_6O + C_2H_4O_2 + H_3N + CO_2 + H_2$	1.28×10^{-02}
20.	Tryptophan	$C_{11}H_{12}N_2O_2 + 2\ H_2O \rightarrow C_8H_7N + C_2H_4O_2 + H_3N + CO_2 + H_2$	1.28×10^{-02}
21.	Glycine	$C_2H_5NO_2 + 0.5\ H_2O \rightarrow 0.75\ C_2H_4O_2 + H_3N + 0.5\ CO_2$	1.28×10^{-02}
22.	Alanine	$C_3H_7NO_2 + 2\ H_2O \rightarrow C_2H_4O_2 + H_3N + CO_2 + 2H_2$	1.28×10^{-02}

(continued)

Table 2 (continued)

No.	Compound	Chemical reactions	Kinetic constant
23.	Cysteine	$C_3H_6NO_2S + 2\ H_2O \rightarrow C_2H_4O_2 + H_3N + CO_2 + 0.5\ H_2 + H_2S$	1.28×10^{-02}
Acidogenic reactions			
24.	Dextrose	$C_6H_{12}O_6 + 0.1115\ H_3N \rightarrow 0.1115\ C_5H_7NO_2 + 0.744\ C_2H_4O_2 + 0.5\ C_3H_6O_2 + 0.4409\ C_4H_8O_2 + 0.6909\ CO_2 + 1.0254\ H_2O$	9.54×10^{-03}
25.	Glycerol	$C_3H_8O_3 + 0.4071\ H_3N + 0.0291\ CO_2 + 0.0005\ H_2 \rightarrow 0.04071\ C_5H_7NO_2 + 0.94185\ C_3H_6O_2 + 1.09308\ H_2O$	1.01×10^{-02}
Acetogenic reactions			
26.	Oleic acid	$C_{18}H_{34}O_2 + 15.2396\ H_2O + 0.2501\ CO_2 + 0.1701\ H_3N \rightarrow 0.1701\ C_5H_7NO_2 + 8.6998\ C_2H_4O_2 + 14.4978\ H_2$	3.64×10^{-12}
27.	Propionic acid	$C_3H_6O_2 + 0.06198\ H_3N + 0.314336\ H_2O \rightarrow 0.06198\ C_5H_7NO_2 + 0.9345\ C_2H_4O_2 + 0.660412\ CH_4 + 0.160688\ CO_2 + 0.00055\ H_2$	1.95×10^{-07}
28.	Isobutyric acid	$C_4H_8O_2 + 0.0653\ H_3N + 0.8038\ H_2O + 0.0006\ H_2 + 0.5543\ CO_2 \rightarrow 0.0653\ C_5H_7NO_2 + 1.8909\ C_2H_4O_2 + 0.446\ CH_4$	5.88×10^{-06}
29.	Isovaleric acid	$C_5H_{10}O_2 + 0.0653\ H_3N + 0.5543\ CO_2 + 0.8044\ H_2O \rightarrow 0.0653\ C_5H_7NO_2 + 0.8912\ C_2H_4O_2 + C_3H_6O_2 + 0.4454\ CH_4 + 0.0006\ H_2$	3.01×10^{-08}
30.	Linoleic acid	$C_{18}H_{32}O_2 + 15.356\ H_2O + 0.482\ CO_2 + 0.1701\ H_3N \rightarrow 0.1701\ C_5H_7NO_2 + 9.02\ C_2H_4O_2 + 10.0723\ H_2$	3.64×10^{-12}
31.	Palmitic acid	$C_{16}H_{34}O + 15.253\ H_2O + 0.482\ CO_2 + 0.1701\ H_3N \rightarrow 0.1701\ C_5H_7NO_2 + 8.4402\ C_2H_4O_2 + 14.9748\ H_2$	3.64×10^{-12}

- There are four streams in this simulation, i.e. INPUT, 5, 6 and LIQUID as shown in Fig. 1. The INPUT stream is the total amount of food waste and its compositions. Stream 6 is the results after the mixer. Stream 5 is the output of the RSTOIC and some input need to be stated to make sure that the data for the calculation block is completed. The LIQUID stream is the final output which is the bioliquid.
- The input streams consist of large components of the food waste as stated in Table 3. The major component of this input stream is water because the reaction in the simulation depends on a high intake of water to produce bioliquid. The input components can be changed into any desired characteristics of waste in handling different types of food waste.

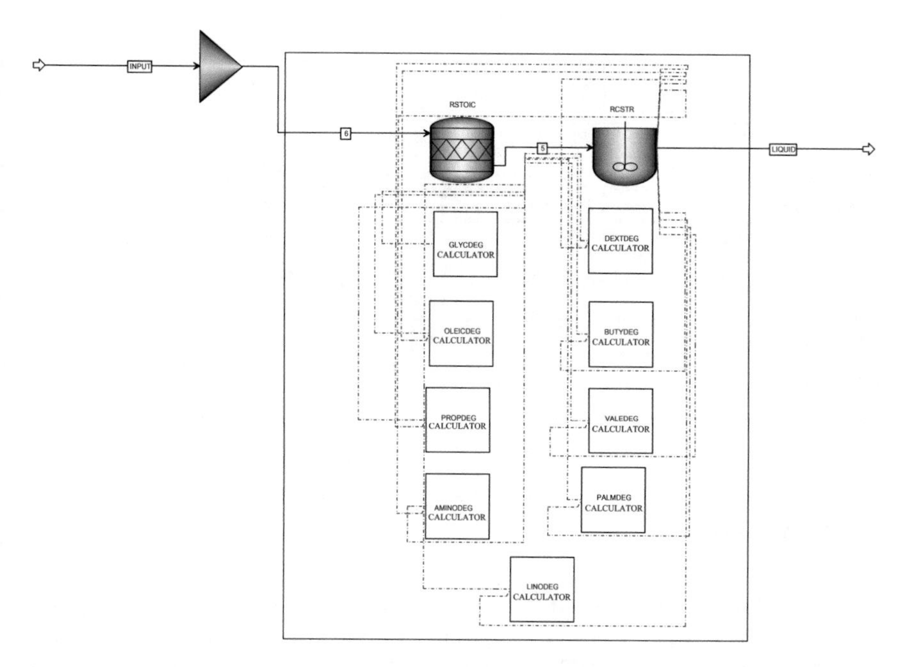

Fig. 1 Flowsheet of anaerobic digestion process model in Aspen Plus

Table 3 The mass fraction of the input stream

Component	Value (kg/kg)	Component	Value (kg/kg)
Water	0.753	Triolein	0.005
Cellulose	0.022	Tripalmitin	0.005
Hemicellulose	0.017	SN-1—01	0.005
Dextrose	0.166	Protein	0.013
NH_3	0.006	Keratin	0.008
Inert	0.002		

- Stream 5 needed to be stated for the calculator block calculation and the value is fixed not like the stream input which can be changed for different types of inputs. Table 4 shows the molar flow rate of the components in stream 5:

Table 4 The input of stream 5 (kmol/h)

Component	Value (kmol/h)	Component	Value (kmol/h)	Component	Value (kmol/h)
Water	50	Phenylalanine	0.039	Alanine	0.177
Glycerol	0.533	Methionine	0.025	Hydrogen	0.500
Oleic acid	0.1	Threonine	0.205	Cellulose	1.882
Dextrose	5.937	Serine	0.589	Hemicellulose	1.882
NH_3	2	Leucine	0.046	Triolein	1.882
CO_2	0.1	Isoleucine	0.059	Tripalmitin	1.882
Arginine	0.074	Valine	0.112	SN-1—01	0.941
Lysine	0.044	Glutamic	1.818	SN-1—02	0.941
Tyrosine	0.017	Aspartic	0.432		
Tryptophan	0.039	Glycine	0.288		

3 Discussion

The simulation results obtained from Aspen Plus are recorded and compared with the experimental data provided by NADNO [16]. Some of the components defined in Aspen Plus are not available in NADNO [16] as shown in Table 5. The comparison is done in terms of the composition of component produced in stream 5. The main component produced is glucose, which is similar with NADNO [16]. The amount of dry matter contained in the bioliquid is 24 wt%. According to NADNO [16], the dry matter content of bioliquid is approximately 20 wt%. It shows that the simulations results obtained from this study show reasonable agreement in comparison with data reported in NADNO [16].

Table 5 Comparison between Aspen Plus simulation and experimental results by NADNO [16]

Component	Mass fraction (wt%)	
	NADNO (2012)	Aspen Plus (this work)
Acetic acid	0.058	0.073
Propionic acid	0.003	0.048
2-Methylpropionic acid	0.001	NA
Butanoic acid	0.011	0.051
Pentanoic acid	0.0002	0.008
3-Methyl butanoic acid	0.001	NA
Cellobiose	0.013	NA
Glucose	**0.509**	**0.229**
Xylose	0.077	0.015
Lactate	0.324	NA
Others	0	0.576

Table 6 Simulation results using Aspen Plus software

	Inlet	Hydrolysis (stream 5)	Bioliquid
Temperature (K)	298.15	328.15	328.15
Pressure (Pa)	101,325	102,300	101,325
Mass flow (kg/day)	48,000	48,000	48,000
Mass fraction			
Water	0.830	0.818	0.813
VFA	0	0.028	0.079
Protein	0.018	0.002	0.002
Keratin	0.017	0.008	0.008
Amino acid	0	0.002	0
NH_3	0.007	0.009	0.01
Sugar	0.092	0.064	0.027
Fat	0.018	0	0
Others	0	0.042	0.039
Inert	0.018	0.018	0.018
Total	1	1	1

Detailed results of the Aspen Plus simulation are summarized in Table 6. The major components of bioliquid are categorized into two groups, for example, sugar and volatile fatty acids (VFAs). Small amounts of sugar such as glucose (defined as dextrose in the simulation) are produced because most of it is converted to acetic acid. Sugars exist predominantly at stream 5 in which the hydrolysis process just occurs. During this process, the large components of food waste have undergone breakdown and form simple monomers of each nutrient, which are glucose, amino acids and fatty acids. The value for simple sugars such as glucose is the highest among all of the other simple monomers that existed in stream 5. These simple sugars continued to the next phase which is the acidogenic and acetogenic reaction that lead to the process of forming more VFA.

VFA is the major component of bioliquid other than water which consists of acetic acid, butyric acid, propionic acid and valeric acid. These components exist in stream 5 which is after hydrolysis, but the VFA content is less because the reaction occurring simply degrades the large components into small monomers, which are sugars, amino acids and fatty acids. The amount of VFA is increased in the final stream after the acidogenic and acetogenic reactions. The large amount of VFA formed at the end of the reaction is reasonable because the further process of the bioliquid leads to the production of fertilizer or biogas. These productions need a bioliquid that has a high content of VFA mainly acetic acid.

In addition, sensitivity analysis is conducted to analyse how the variation of residence time would affect the production of bioliquid. The range of residence times tested was in the range 12–20 h. Figure 2 shows the production of bioliquid at various residence times. The amount of bioliquid produced is increased slightly as residence time increased. A 5% increment in bioliquid production is observed as

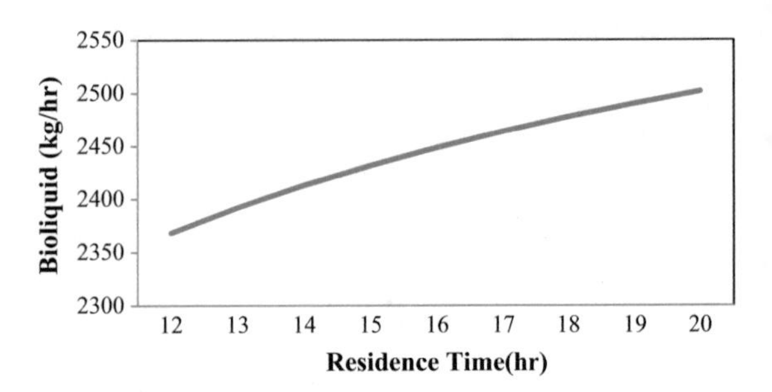

Fig. 2 Bioliquid production at various residence times

the residence time increased from 12 to 20 h. The optimum residence time depends on the load capacity and type of enzymes used in the reactor [16]. The longer residence time will allow the enzymatic process to be more efficient where the bonds of the organic materials will be broken and hydrolysis will occur [17].

4 Conclusion

The anaerobic digestion process for production of bioliquid from MSW was simulated using Aspen Plus software. The amount of dry matter produced in bioliquid was 20 wt% which consists mainly of VFA. Bioliquid with a large amount of VFA can be further utilized to produce biogas. A sensitivity analysis conducted at different residence times shows a slight increase in bioliquid as the residence time increases. The model developed can be used for estimating the production of bioliquid at various operating conditions.

References

1. Noor ZZ, Yusuf RO, Abba AH, Abu Hassan MA, Mohd Din MF (2013) An overview for energy recovery from municipal solid wastes (MSW) in Malaysia scenario. Renew Sustain Energy Rev 20:378–384
2. Malaysia Second National Communication to the UNFCCC (2007) ISBN 978-983-44294-9-2
3. Agamuthu P, Fauziah SH, Kahlil K (2009) Evolution of solid waste management in Malaysia: impacts and implications of the solid waste bill, 2007. Mater Cycles Waste Manage 11(2): 96–103
4. Tan ST, Ho WS, Hashim H, Lee CT, Taib MR, Ho CS (2015) Energy, economic and environmental (3E) analysis of waste-to-energy (WTE) strategies for municipal solid waste (MSW) management in Malaysia. Energy Convers Manage 102:111–120
5. Fazeli A, Bakhtvar F, Jahanshaloo L, Sidik NAC, Bayat AE (2016) Malaysia's stand on municipal solid waste conversion to energy: a review. Renew Sustain Energy Rev 58:1007–1016
6. Mekhilef S, Saidur R, Safari A, Mustaffa WESB (2011) Biomass energy in Malaysia: current state and prospects. Renew Sustain Energy Rev 15(7):3360–3370

7. Nguyen HH (2014) Modelling of food waste digestion using ADM1 integrated with Aspen Plus. University of Southampton, Southampton
8. Murto M, Bjornsson L, Mattiasson B (2004) Impact of food industrial waste on anaerobic co-digestion of sewage sludge and pig manure. J Environ Manage 70:101–107
9. Banks C, Zhang Y (2010) Technical report: optimising inputs and outputs from anaerobic digestion processes. Defra Project Code WR0212. 2010
10. Moeller L, Goersch K, Neuhaus J, Zehnsdorf A, Mueller R (2012) Comparative review of foam formation in biogas plants and ruminant bloat. Energy, Sustain Soc 2:1–9
11. Suhartini S (2014) The anaerobic digestion of sugar beet pulp. University of Southampton, UK
12. Donoso-Bravo A, Mailier J, Mailier J, Martin C, Rodriguez J, Aceves-Lara CA, Wouwer AV (2011) Model selection, identification and validation in anaerobic digestion: a review. Water Res 45:5347–5364
13. Batstone DJ, Keller J, Angelidaki I, Kalyuzhnyi SV, Pavlostathis SG, Rozzi A, Sanders WTM, Siegrist H, Vavilin VA (2002) The IWA anaerobic digestion model no 1 (ADM 1). Water Sci Technol 45(10):65–73
14. Serrano RP (2011) Biogas process simulation using Aspen Plus®. Master thesis, Syddansk Universitet, Denmark
15. Rajendran K, Kankanala HR, Lundin M, Taherzadeh MJ (2014) A novel process simulation model for anaerobic digestion using Aspen Plus. Bioresour Technol 168:7–13
16. Nanna Dreyer Nørholm (NADNO) (2012) REnescience PSO report 7335. http://www.energinet.dk/SiteCollectionDocuments/Danske%20dokumenter/Forskning%20-%20PSO-projekter/7335%20Renescience-PSO%20report%20Final.pdf. Accessed May 2016
17. Tonini D, Dorini G, Astrup TF (2014) Bioenergy, material, and nutrients recovery from household waste: advanced material, substance, energy, and cost flow analysis of a waste refinery process. Appl Energy 121:64–78

Anaerobic Digestion of Screenings for Biogas Recovery

N. Wid and N. J. Horan

Abstract Screenings comprise untreatable solid materials that have found their way into the sewer. They are removed during preliminary treatment at the inlet work of any wastewater treatment process using a unit operation termed as a screen and at present are disposed of to landfill. These materials, if not removed, will damage mechanical equipment due to its heterogeneity and reduce overall treatment process, reliability and effectiveness. That is why this material is retained and prevented from entering the treatment system before finally being disposed of. The amount of biodegradable organic matter in screenings often exceeds the upper limit and emits a significant amount of greenhouse gases during biodegradation on landfill. Nutrient release can cause a serious problem of eutrophication phenomena in receiving waters and a deterioration of water quality. Disposal of screenings on landfill also can cause odour problem due to putrescible nature of some of the solid material. In view of the high organic content of screenings, anaerobic digestion method may not only offer the potential for energy recovery but also nutrient. In this study, the anaerobic digestion was performed for 30 days, at controlled pH and temperature, using different dry solids concentrations of screenings to study the potential of biogas recovery in the form of methane. It was found screenings have physical characteristics of 30% total solids and 93% volatile solids, suggesting screenings are a type of waste with high dry solids and organic contents. Consistent pH around pH 6.22 indicates anaerobic digestion of screenings needs minimum pH correction. The biomethane potential tests demonstrated screenings were amenable to anaerobic digestion with methane yield of 355 m^3/kg VS, which is comparable to the previous results. This study shows that anaerobic digestion is not only beneficial for waste treatment but also to turn waste into useful resources.

N. Wid (✉)
Faculty of Science and Natural Resources, Universiti Malaysia Sabah, Jalan UMS, 88400 Kota Kinabalu, Sabah, Malaysia
e-mail: newati@ums.edu.my

N. J. Horan
School of Civil Engineering, Faculty of Engineering, University of Leeds, Leeds, West Yorkshire LS2 9JT, UK

N. Horan et al. (eds.), *Anaerobic Digestion Processes*,
Green Energy and Technology, https://doi.org/10.1007/978-981-10-8129-3_6

Keywords Anaerobic digestion · Biogas recovery · Screenings

1 Introduction

Biogas is a renewable energy primarily consists of a mixture of methane (CH_4) and carbon dioxide (CO_2) with a small amount of other gases. Biogas can be recovered from organic wastes, which received intensive research attention these decades [1]. It is produced through a breakdown of the organic matter in the absence of oxygen. This process is known as an anaerobic digestion. Biogas is considered as a renewable energy source because of the suitability as an energy production due to the high content of the methane gas [2]. The world main energy consumption such as crude oil, coal and natural oil is reported to be diminished in future, which increases energy prices, mainly due to increasing world population, economy and industrial and agricultural [3]. More than 80% of the global energy consumption is supplied by fossil energy sources, which are quickly being depleted. In order to replace and reduce the amount of the fossil energy sources used, biogas is the best compound that can act as an energy source to replace the depleting fossil fuels. Biogas that derived from biological sources can reduce the dependence on these depleting natural resources and address the energy insecurity concerns due to its renewable, widely applicable with various advantages [4]. Anaerobic digestion offers a complete package which provides a sustainable approach that combines waste treatment with a recovery of renewable energy and nutrient [3, 5, 6]. The biological technique also consumes minimum energy with lower carbon footprints and reduces organic waste into stabilised end products. Screenings comprise untreatable solid materials that removed during preliminary treatment of wastewater treatment process and disposed of to landfill. This material contains high biodegradable organic matter that often exceeds the upper limit and emits a significant amount of greenhouse gases (GHGs) during biodegradation. Therefore, a study is performed to convert this organic waste into energy by using anaerobic digestion.

2 Screenings

Screenings are produced during wastewater treatment and they are considered as municipal solid waste as they occur predominantly as coarse solids, and at present are disposed of to landfill. The first stage in wastewater treatment is to remove untreatable solid material such as rags, paper, plastic, wood and other material that has found its way into the sewer. It is achieved using a unit operation termed a screen, and consequently, the material removed is termed screenings. These materials, if not removed, will damage mechanical equipment such as pumps, aerators and other process equipment, block pipes and valves, contaminate waterways and reduce overall treatment process, reliability and effectiveness [7–9]. Due to the relatively low production of screenings as compared to sludge production, little attention has been paid so far to this type of waste [10].

2.1 Handling and Disposal of Screenings

The screenings removed during wastewater treatment apart from the solid contents have high water content and organic content. In conventional treatment, screenings are removed by screens, after which they may be further treated for use as a fertiliser, or disposed of. Due to the putrescible nature of some of the solid materials, there will be a problem of odour if careful thought is not given to their handling and disposal. Consequently, they are generally subjected to further treatment before disposal. Thus, once they are removed from the flow, they undergo handling involving macerating and compaction. During this process, the screenings are dewatered and partially compacted before passing to skip prior to final disposal. At this stage, much of the faecal materials are washed out from the screenings. Compacted screenings can have dry solids of up to 40% and the conditioning stage can achieve volume and weight reductions as high as 70% [9]. Figure 1 shows compacted screenings at Knostrop WWTP, in Leeds. Every year, around 150,000 tonnes of screenings and grit are produced in the UK [11].

Fig. 1 Compacted screenings in a skip before final disposal [9]

In Europe, the most commonly used disposal techniques for screenings in Europe are landfilling and incineration [8, 12]. However, landfilling is not favoured by the European waste regulations as the EU Landfill Directive requires: (i) a reduction in the amounts of biodegradable organic matter disposed into landfills, and (ii) a ban on waste with a water content above 70% w/w to be landfilled. Screenings often exceed this upper limit [8, 10, 13, 14]. Incineration is considered as a good alternative; however, the high moisture content may jeopardise the operating conditions of the incineration plant [12]. With technological evolutions in wastewater treatment processes that require fine sieving or pretreatment, the production of screenings is expected to increase in coming years. Therefore, alternative treatments that prove to be more adapted for screenings need to be developed prior to final disposal.

2.2 Characteristics and Composition of Screenings

Very few studies have been carried out on screening materials. The few that have been undertaken have been more interested in the technical establishment of the treatment processes and composition and characterisation of the materials. So far, reports on the potential for resource recovery from screenings are very rare.

Screenings consist of untreated domestic solid materials that removed during wastewater treatment works. Screenings are considered with six significant fractions produced through the screen, i.e. sanitary textiles, fine fractions, vegetable, papers and cardboards, plastics and 'others' (comprising all the other types of materials). Le Hyaric et al. [15] investigated the development of disposal strategies for screenings at minimal costs, within Severn Trent Water, UK. This study reported data relative to the volumes of screenings generated, their composition and the existing treatment methods. They identified two methods to assess maximum production rate of screenings, i.e. determine the number of screenings-loaded containers or 'skip' emptied per year, and record scientifically the screening produced, which is relatively expensive. In terms of the material dryness, they adopted at least 25% dry solids of the material produced from the handling equipment, to avoid spillage during transport and acceptable for landfill. The approach of reducing the material size to fine pieces that can be returned to sewage flow and removed from the plant with sludge is not favoured in the treatment works because a maceration unit tends to lose efficiency quite quickly and causes problems on bacteria beds, with sludge processing equipment and with the use of sludge on farmland. They also suggested covering the storage container to prevent rainwater accumulation which can cause problems in transportation and disposal. But these studies were carried out in 1996, and since then, consumer habits, wastewater collection systems and treatment technologies have changed. Consequently, screenings composition may highly affected. Table 1 compares the compositions of screenings as reported by different studies which were carried out in 1996 and 2016.

Table 1 Composition of typical screenings arriving at domestic treatment works

Component	Dry weight (%)	
	Reference [14]	Reference [15]
Paper	20–50	62–67
Rags	15–30	23–32
Plastic	5–20	2–3
Rubber	0–5	–
Vegetable matter	0–5	5–6
Faecal matter	0–5	0–2

Nowadays, with the technology evolution of municipal wastewater treatment processes, the production of screenings is expected to increase as the gap sizes of the screened used have decreased from a few cm to a few mm. In some wastewater treatment plants, the gap size can bear as low as 1 mm where membrane bioreactors are used [10]. Le Hyaric et al. [15] evaluated the influence of gap size of the screens used, compaction of the waste and the weather conditions on the quantity and quality of screenings. When a low gap is used, the fraction of sanitary textiles or other solids materials with bigger size is found to decrease with the gap size. Compaction treatment also affects the mass of waste generated through dewatering process, and the dry season produced less waste compared to the wet season.

Wid and Horan [16] characterised the screenings composition which collected from the Knostrop Wastewater Treatment Plant, located in Leeds, UK. As seen in Table 2, the composition of paper and plastic shows a significant difference after 20 years of the first study. These are the common two types of materials used domestically. The high composition of paper suggests current practice in using paper products, such as paper bags and cardboards as packaging materials. It has

Table 2 Single-phase (conventional system) performances from previous studies

Substrate	Methane yield (mL CH_4/g VS_{added})	Methane content (%)	References
Fruits and crops	500–600	51–53	[42]
MSW	140	–	[19]
French bean waste	470	71–75	[46]
Bermuda grass	112–219	22.6–41.3	[47]
Sweet sorghum cultivar	360	36	[48]
Source separated OFMSW	398	62.5	[18]
Mixed primary and activated sludges	225	70.3	[33]
Terrestrial weeds	117	73	[49]
Terrestrial weeds	115	72	[49]
Vegetable waste	400	64	[50]
Poultry wastes	90 550	– –	[51]

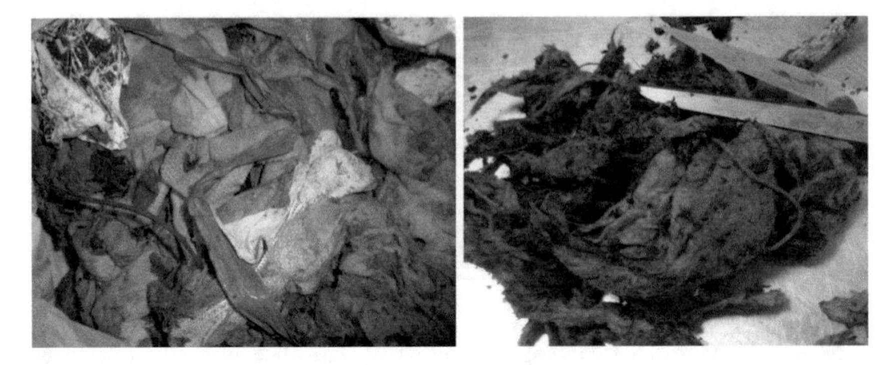

Fig. 2 Sanitary products found in the screenings [9]

been noted that paperboard and paper are always the largest component of the municipal solid waste stream [17]. According to [18], a high paper composition results in higher volatile solids, i.e. above 82%, consequently higher biodegradability and biogas. This finding also opens the possibility for screenings to be recycled from waste stream, especially paper recycling. While reduction in plastics generation may due to the gradual public awareness on the drawbacks of using plastic materials such as low biodegradability and negative impacts to the environments. The composition of rags shows consistent generation. The high proportion of rags underlined the increasing use of disposable wipes [15, 16]. It can be seen from Fig. 2 that sanitary towels and tampons, which are mainly made of cotton, comprise a large fraction of the rags in screenings. This large fraction of sanitary material is another reason why screenings are categorised as a difficult waste, which may lead to difficulties in storing and disposal [8].

3 Anaerobic Digestion Process

Anaerobic digestion (AD) is a degradation process of organic matter by specific microorganisms in the absence of oxygen. Anaerobic digestion is also known as an environmentally sustainable technology for converting a variety of waste resources including organic fraction of municipal solid waste, manure and agricultural residue to energy in the form of methane and hydrogen. During wastewater treatment process, anaerobic digestion is used to treat or stabilise sludge before it is disposed or recycled.

Among biological treatments, anaerobic digestion is reported as the most cost-effective, due to the high energy recovery and less environmental impact [19] and commonly used at wastewater treatment plants to degrade sludge [20]. The main products of anaerobic digestion are biogas and digestate, i.e. stabilised solid and liquid residue of digestion. Biogas consists mostly of methane and carbon dioxide [21, 22]. The final products of anaerobic digestion are not only reduced in mass and stabilised but also the biogas produced can be further used as a source of fuel, heat and electricity, and the digestate can be used as soil conditioner to fertilise land. The liquid part from AD can be further used to precipitate struvite, a slow-release fertiliser [16, 23].

The microbiology of anaerobic digestion involves several complicated processes but four main steps are distinguished, i.e. hydrolysis, fermentation, acetogenesis and methanogenesis, and involves three major bacterial groups, i.e. hydrolytic-fermentative bacteria, acetogenic bacteria and methanogenic bacteria. These bacteria are responsible for converting organic compounds to volatile fatty acids (VFAs) with the simultaneous production of hydrogen and carbon dioxide, converting the acids to acetic acid and produce methane either from acetate or hydrogen or carbon dioxide, respectively [24]. The nature of the anaerobic substrate chain is such that each of these groups relies on the previous one for its substrate and passes the products onto the next one to avoid product accumulation [20]. Figure 3 shows the chemical pathways followed during the conversion of complex organic material to methane, as described by [25, 26].

3.1 Hydrolysis

Hydrolysis means both the solubilisation of insoluble particulate matter and the biological decomposition of particulate complex organic matter (biopolymers proteins, carbohydrates, lipids) into simples molecules (monomers or dimers),

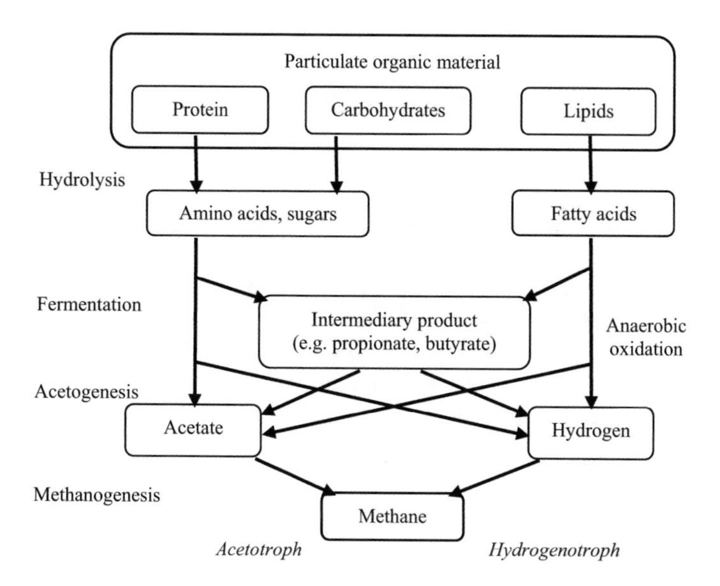

Fig. 3 Reaction scheme for anaerobic digestion of domestic sewage sludge [25, 26]

which can pass the cell membrane. The process is carried out by extracellular enzymes (hydrolases) and it may or may not be the rate-limiting step of their bioconversion under anaerobic digestion [27]. Hydrolysis products include sugars, amino acids, long chain fatty acids (LCFA) and glycerol [24, 28].

3.2 Fermentation

Fermentation is the breakdown of soluble materials produced by the hydrolysis process. The process is also known as acidogenesis as the main products are volatile fatty acids (VFAs) (e.g. acetic, propionic, butyric and other volatile acids) which are converted by acid-forming bacteria (acid formers). The soluble materials either fermented directly into acetic acid or into intermediate products, such as propionate and butyrate, which will be further fermented into acetate. These organic acids become the substrate for the next steps in acetogenesis and methanogenesis.

In this stage, acidogenesis is generally considered to be the fastest of the individual steps in the anaerobic process [28], means that in case of overloading of organic acids in the reactor, and causes a drop in pH, decrease in alkalinity and finally a failure of the digester. A syntropic relationship exists between acid-forming organisms and acid-utilising organisms such as methane-forming bacteria to balance the digester performance [29].

$$4H_2 + 2CO_2 \rightarrow CH_3COOH + 2H_2O \quad (1)$$

$$4CO + 2H_2O \rightarrow CH_3COOH + 2CO_2 \quad (2)$$

$$4CH_3OH + CO_2 \rightarrow 3CH_3COOH + 2H_2O \quad (3)$$

$$C_6H_{12}O_6 \rightarrow 3CH_3COOH \quad (4)$$

$$HCOOH \rightarrow H_2 + CO_2 \quad (5)$$

In this stage, intermediate products from the fermentation process are converted to acetate by acetogenic-forming bacteria, i.e. homoacetogens. Most acetogenic bacteria produce acetate from hydrogen (H_2) and carbon dioxide (CO_2) (1), while some produce acetate from water (H_2O) and carbon monoxide (CO) (2), some from carbon dioxide (CO_2) and methanol (CH_3OH) (3), and often six-carbon sugars or hexoses are degraded to acetate (4). Hydrogen is also produced from long and short-chain fatty acids, for example, formate decomposition to H_2 and CO_2 (5).

3.3 Methanogenesis

The methanogenic phase is normally considered the limiting step of metabolic reactions in anaerobic digestion due to the slow growth rate of the methanogenic bacteria. Hydrolysis can be the controlling step in the conversion process if the substrate was particulate and comprises predominantly cellulose [30].

The methanogens rely on the acetogens to provide them with acetate, hydrogen and carbonate, while the acetogens rely on the methanogens to remove hydrogen [22]. Methanogens are grouped as acetoclastic (6) and hydrogenotrophic (7), which can produce methane (CH_4) 70–75% and 25–30%, respectively, according to the substrate that they can utilise, where acetate is the major substrate used by the methane-forming bacteria.

Acetoclastic methanogenesis (70–75%):

$$CH_3COOH \rightarrow CH_4 + CO_2 \tag{6}$$

Hydrogenotrophic methanogenesis (25–30%):

$$CO_2 + 4H_2 \rightarrow CH_4 + 2H_2O \tag{7}$$

Methanogenesis is considered the most sensitive stage in anaerobic digestion. The methanogenic bacteria are highly sensitive to pH fluctuations, temperature, loading rate and they are inhibited by a number of compounds as reported in previous studies [27].

4 Type of Anaerobic Digester

Anaerobic digestion of waste can be performed using different reactor systems, which also known as anaerobic digester. For example, single-phase, two-phase or multiphase configuration.

4.1 Single-Stage Digester

Single-stage digesters consist of only one reactor and operations consist of waste substrate feeding and withdraw, mixing, heating and gas collection [20]. Usually, the digesters are fed intermittently with once a day is the most common period of feeding rate. In a continuous stirred tank reactor (CSTR), an influent substrate concentration of 3–8% total solids (TS) is added daily and an equal amount of effluent is withdrawn. The digester is maintained constantly either under mesophilic or thermophilic condition [9, 31].

Single-stage digesters are more easily upset than two-stage digesters [20]. This is because different groups of microorganisms develop in the same environment, i.e. acid formers and the methane formers. In a well-balanced digester, each group of bacteria will establish its own particular population which, in turn, depends on the feed material, operating conditions (pH, temperature, retention period) and on the stoichiometry of the reaction involved [32]. If acid formers grow more quickly than methane formers, an imbalance between acid production rate and methane production rate often occurs. This imbalance may cause a decrease in alkalinity and pH that results in digester failure [29]. The problems of imbalance can be obviated by operating a single-stage digester at low hydraulic retention times (HRTs) and low organic loading rates (OLRs) [33]. The best way to overcome the inhibitory effects of intermediate products produced during the early stages of digestion is by separating the process into two stages [29, 33].

4.2 Two-Stage Digester (Two-Phase)

These systems improve efficiency and stability over a single-stage system which carries out the same operations as the single-stage. The two-phase digestion system consists of two separate biochemical reactors in series, where fermentation and methanogenesis are physically separated. The first phase is referred to as 'acid fermentation' and involves the production of volatile fatty acids (VFAs), while the second phase is referred to as 'methane fermentation' because in it the VFAs are converted to methane and carbon dioxide [34] (Fig. 4). Because the volatile acids are primary products of the first phase, pH control in the second phase may be necessary when the buffer capacity has been exceeded. Such control in the methane reactor could be provided by external neutralisation of the influent or by recycling of supernatant from the second phase [9, 33].

The benefits of two-phase system including to reduce considerable total digestion time than the conventional single-stage digestion consequently reduced the reactor size and capital cost, improve the effluent quality, methane yield, volatile

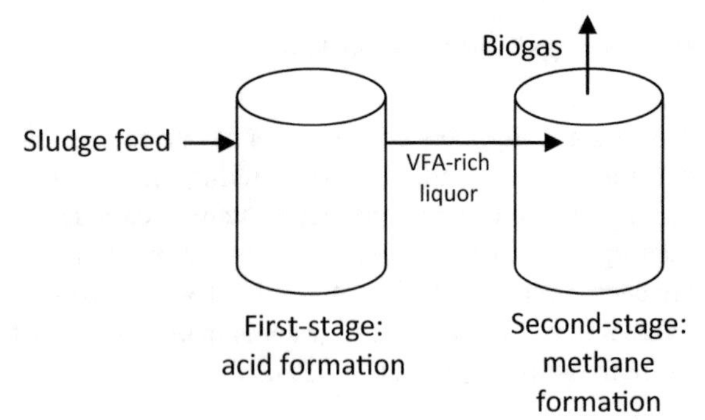

Fig. 4 Two-phase digester [9]

solid reduction and process stability [34]. This implies that the performance of an anaerobic process could be improved with the proper combination of the anaerobic process characteristics. Various types of reactors have been used in the two-stage anaerobic digestion process, for example, leaching bed reactor (LBR) for acidogenesis, upflow anaerobic sludge blanket (UASB) for methanogenesis and continuous stirred tank reactor (CSTR) in both systems.

5 AD, Its Role in Biogas Recovery

The world energy demand is expected to grow by as much as 50% by 2025, mainly due to the increasing world population, economy, industrial and agricultural, leading to the consumption of more fossil fuels with increasing energy prices [3]. This will bring about a rapid depletion of fossil fuels reserves. Therefore, in order to deal with the large quantities of waste materials disposed to landfills, with the GHG emissions, it is apparent that intense implementation needs to be performed. Anaerobic digestion offers a complete package which provides a sustainable approach that combines waste treatment with a recovery of renewable energy [3] as well as nutrient [35, 36]. Anaerobic digestion technique offers significant advantages including consume minimum energy with a lower carbon footprint [24] and reduce organic waste into stabilised end products [16], generating renewable energy and a supply of digested materials (biosolids or digestate) as a fertiliser or soil conditioner for crop production. In the UK, a number of government subsidies have been introduced to promote renewable energy and heat generated by anaerobic digestion such as Renewable Obligations Certificate (ROC), Feed-In Tariffs Schemes (FITS) and Renewable Heat Intensive (RHI) [9].

In anaerobic digestion study, special attention has been paid to the final stage of the process, i.e. methanogenesis, as this is considered as the most important step of the overall process and there are many successful attempts reported in the literature regarding of digestion of wastewater to produce methane (Table 2).

5.1 Biochemical Methane Potential (BMP)

The biochemical methane potential (BMP) assay is developed by [37] to determine the ultimate methane yield of organic substrate. It offers the most promise for resolving anaerobic treatment problems because it is relatively simple, quick and inexpensive. The technique is very useful for sorting out important variables and for the development of an efficient continuous-feed assay programme. Anaerobic batch studies using the BMP assay are performed for three purposes; anaerobic biodegradability, inoculum activity and inhibition, based on biogas production [38].

The basic approach of BMP is to incubate a small amount of waste (substrate) with an anaerobic inoculum and appropriate nutrients at a controlled temperature.

Fig. 5 Laboratory scale 400 mL anaerobic reactors used in BMP testing [9]

Normally, mesophilic digestion is carried out at temperature ranges from 20 to 45 °C and for thermophilic ranges from 46 to 60 °C [38]. The pH is maintained in the range of 6.8–7.2 throughout 30 days digestion period. For laboratory scale, the assay is performed in closed bottles of 100 mL up to 2 L, depending on the homogeneity of the substrate (Fig. 5). Typically, the assay vessels are flushed using nitrogen (N_2) prior to digestion. These should be continuously stirred and kept under anaerobic conditions during the process of transfer. The assay bottles can be placed in a mixing incubator at 100 rpm [39], or can be mixed manually on each bottle every one or two days during the entire incubation period [40]. The batches are occasionally shaken and moved around in the incubator to compensate for any minor variation in temperature in different parts of the incubator [41].

The measurement of the biogas produced was performed using liquid displacement method to scrub CO_2. NaOH with concentration 5% was used and the volume of biogas produced was corrected to standard temperature (0 °C) and pressure (1 atm). To ensure complete biodegradation of the material, the background values for 30 days of digestion was subtracted [16]. The methane potential can be determined in mL CH_4 per gramme of organic waste expressed as volatile solids added (mL CH_4/g VS_{added}) or removed (mL CH_4/g $VS_{removed}$). Some studies express methane recovery using CH_4 per gramme of total solids added (mL CH_4/g TS_{added}) or CH_4 per gramme of total solids removed (mL CH_4/g $TS_{removed}$).

5.2 *Methane Phase Digestion (MPD)*

Anaerobic digestion needs a long time for the digestion to complete and its decomposition efficiency is low. Methane phase digestion or methanogenesis is the last stage in the anaerobic digestion process, which the predominant biogas produced is methane. Methanogenesis is highly depended on temperature, pH and many other parameters; therefore, meticulous attention needs to be paid when performing the digestion process. Methanogenesis can be carried out using single and two-stage reactors. When using continuous reactor such as CSTR, other factors such as HRT and OLR are very important. In general, the HRT for anaerobic digestion of solid waste usually ranges between 16 and 20 days (d) when a single-reactor is employed [42]. But by applying two-phase systems, the phases of digestion are separated and the process operates at a shorter HRT with a considerable increase in system stability and reliability. An optimum HRT of 12 d was observed in a methane digester following acid digestion, to produce high methane production rate [43], compared to a single-phase digester which achieved at higher HRT, i.e. 21 d [19]. When single-phase digester operated at 15 d HRT, methane yield and methane content can achieve 225 mL CH_4/g VS_{added} and 70.3%, respectively. But when a two-phase system was employed, higher gas production was obtained at a shorter HRT, i.e. 7 d (302 mL CH_4/g VS_{added} and 64.7%, respectively), albeit with a slightly lower methane content [33]. This suggests in single-stage CSTR digestion, reliable system operation can be expected only at high HRTs, where the rates of acids production and conversion are balanced. This can attribute first to the characteristics of the substrates used and also the operating conditions maintained in the studies. Acclimatisation is also an important factor in methane phase digestion. Acclimatised conditions can be achieved when less than 10% difference between biogas production and pH are observed. For difficult type of substrates such as heterogeneous solid wastes, acclimatisation phase takes longer about two (2) months or more before starting the actual experiments, while for liquid type of substrate such a wastewater effluent, a shorter time is needed.

6 Biogas Recovery from Screenings Using Anaerobic Digestion

6.1 Introduction

The recovery of biogas from organic waste has received considerable attention over the past decades. However, biogas recovery from screenings, a difficult solid material, has received little attention and not been explored widely. It is reported that global resources are depleting; therefore, wastes including screenings are no longer viewed as a waste stream, but as a resource to recover including methane as renewable energy. The most common disposal technique of screenings is by landfilling, which results in GHGs emissions, odour problems and other environmental issues, with lost energy, nutrients and other recoverable resources. Additionally, limited space remaining may jeopardise landfill application. There are different methods employed for the treatment of solid wastes, which one of them is anaerobic digestion. It is a sustainable approach that combines wastewater treatment with the recovery of valuable by-products, including renewable energy and nutrient

[3]. Most of the results discussed in this current study were published in [16] and reported in [9] (N. Wid and N. J. Horan, "Anaerobic digestion of wastewater screenings for resource recovery and waste reduction," *IOP Conference Series: Earth and Environmental Science*, vol. 36 (1), pp. 1–7, 2016; N. Wid, "*Resource Recovery from Screenings through Anaerobic Digestion*," Ph.D., School of Civil Engineering, Faculty of Engineering, University of Leeds, United Kingdom, 2012).

6.2 *Materials and Methods*

6.2.1 Sample Collection and Preparation

The screenings and primary sludge used in this study were collected from a wastewater treatment plant, located in West Yorkshire, Leeds, UK. The sample was simply collected when needed with four major sampling regimes were undertaken. Primary sludge was used as seed inoculum to boost up the digestion process. To achieve a homogenised and uniform size of sample (Fig. 6), the collected screenings were cut and ground using a scissor and grinder, respectively, after removing plastics. The prepared sample was kept in a fridge at 4 °C prior to use.

6.2.2 Characterisation of Screenings

The fresh screenings were characterised in terms of the physical properties including total solids (TS), volatile solids (VS) and pH. The analyses were performed according to the Standard Methods [44].

Fig. 6 Homogenised screenings [9]

6.2.3 Anaerobic Digestion

Anaerobic digestion was carried out at a laboratory scale based on the method described by [22, 37, 41]. Anaerobic batch reactors are developed using serum bottles with effective volume of 400 mL, which are sealed and tightened with neoprene bungs. Different dry solids concentrations were studied, 3, 6, 9 and 12%, in four (4) reactors, and one (1) reactor representing the control, which contained only inoculums and nutrient, to correct the biogas production of the inoculum. The ratio of screenings to inoculums was 1.5:1, and nutrient was added at 1 mL/L containing all the micronutrients essential to the growth of the anaerobic microorganisms. Nitrogen gas was sparged to create anaerobic conditions in the reactors before starting the experiments. The experiment was carried out in a shaker incubator with 100 rpm for 30 days at controlled pH (6.8–7.2) and temperature (37 °C). Biogas produced from the reactors was measured using a liquid displacement method [16].

6.3 Results and Discussion

6.3.1 Physical Characterisation of Screenings

To perform anaerobic digestion, the composition of the substrate, in this case screenings, needs to be determined prior to feeding into the digester, to permit modifications to the key parameters. As shown in Table 3, the composition of each sample was very consistent, despite the wide gaps between collection periods. The screenings were sampled at different collection times to obtain representative sample. The average total solids content was 29.6%, which suggests screenings are a difficult solid waste with high dry solids content. According to [15], dewatering the screenings by compaction during preliminary treatment at the plant increases the dry solids content of the sample. Similar results were reported by Huber et al. [13] with 24–30% and Le Hyaric et al. [12] with 30% of total solids. It is suggested the screenings should be at least 25% of dry solids to reduce the cost of disposal and to facilitate waste transportation. Despite the dryness, screenings are still high in moisture and not recommended for landfill. The volatile solids were always higher than 90%, which suggests a high organic matter content. [15] reported similar observations, with volatile solids of screenings were about 90%. High volatile solids illustrate the potential of methane production as well as nutrient recovery. The average pH was 6.2 suggesting that screenings were well buffered throughout the 30 days of digestion without external pH correction, thus provides an ideal feedstock for anaerobic digestion.

Table 3 Physical characterisation of screenings [16]

Physical characteristics	Average (std. dev.)
TS (% dry solids)	29.6 (3.0)
VS (% total solids)	93.3 (1.3)
pH	6.2 (0.3)

Note TS = total solids, VS = volatile solids

6.3.2 Biogas Recovery

Figure 7 shows the profile of cumulative biogas production (GP) at standard temperature and pressure (STP) throughout 30 days of digestion using different dry solids concentrations of screenings. There was negligible difference in biogas production over the first 20 days between feed total solids of 3 and 6%. However, the 6% total solids continued to produce biogas after this period whereas there was a plateau for the 3% solids. Thus, the cumulative biogas production for 6 and 3% reached 4466 mL and 3097 mL of biogas, respectively. In the case of 9 and 12%, the biogas production was poor with a cumulative biogas production of 835 and 756 mL, respectively. This indicates that 6% was the ideal concentration of dry solids for anaerobic digestion, followed by 3%. However, higher concentrations of 9 and 12% were inhibitory to methanogenesis possibly due to overloaded with high volatile fatty acids (VFAs) concentrations or ammonia [16].

In this study, the biogas yield is expressed in terms of volume of gas produced/amount of volatile solids applied (Table 4). This is clearly seen for the digestion of screenings at 6%, where hydrolysis proceeds rapidly over the first 8 days and a rapid increase after 12 days indicates more complex material is destroyed, thus produced 0.36 m^3CH_4/kg VS methane yield. Previous studies also found the yield obtained from screenings is typical of the value of other readily biodegradable

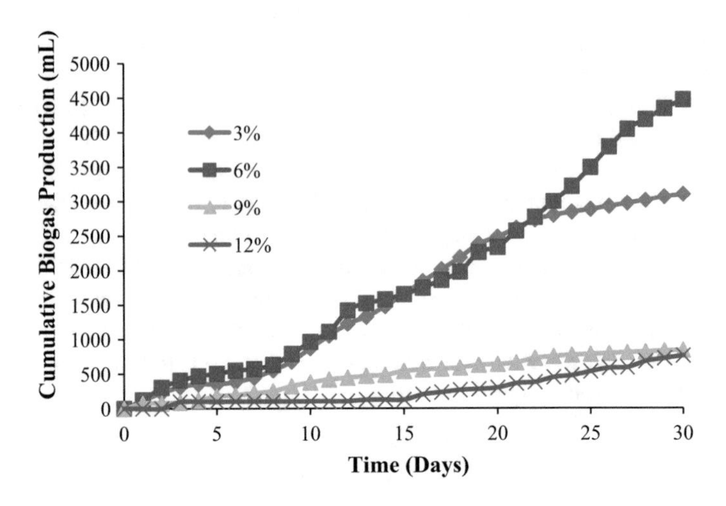

Fig. 7 Cumulative biogas production at different total solids concentrations [16]

Table 4 The effect of dry solids on the methane yield [16]

Reactor (% total solids)	Total biogas production (mL)	Yield (m^3 CH_4/kg VS)
3	3097.11	0.30
6	4465.74	0.36
9	835.21	0.45
12	756.57	0.28

Note The biogas is assumed to be 70% in the form of methane from the total biogas

wastes, such as organic fraction municipal solid waste (OFMSW) with 0.35 m^3CH_4/kg VS [6], food waste with 0.39 m^3CH_4/kg VS [40] and complex organic substrate with 0.24 m^3CH_4/kg VS methane yields [45]. Other studies on methane recovery as stated in Table 2.

6.4 Conclusion

Screenings, heterogeneous solid materials, were investigated for its potential in recovering biogas in the form of methane. The results show that screenings were high in organic content with total solids (also termed as dry solids) and volatile solids contents were 29.6 and 93.3%, respectively. The digestion was well buffered without the need for external pH control as the pH of the material was 6.2 close to the ideal pH for anaerobic digestion. The promising results of methane yield of 0.36 m^3/kg VS suggest that screenings were a potential source for energy production. The ideal dry solids concentration for methane production was 6%. Total solids concentrations higher than 6% were inhibitory to methanogenesis possibly as a result of solids overloading. Even though screenings were categorised as difficult waste due to the large portion of solid material and high total solids content, the high organic content as indicated by volatile solids content of screenings suggests that anaerobic digestion method may not only offer the potential for methane recovery but also nutrient, such as nitrogen and phosphorus. The well-buffered system indicates screenings are amenable to anaerobic digestion with no meticulous attention needed for pH control throughout digestion process.

References

1. Morero B, Groppelli E, Campanella EA (2015) Life cycle assessment of biomethane use in Argentina. Bioresour Technol 182:208–216
2. Sitorus B, Sukandar, Panjaitan SD (2013) Biogas recovery from anaerobic digestion process of mixed fruit-vegetable wastes. Energy Procedia 32:176–182
3. Khanal SK (2008) Anaerobic biotechnology for bioenergy production principles and applications. Wiley, USA

4. Mao C, Feng Y, Wang X, Ren G (2015) Review on research achievements of biogas from anaerobic digestion. Renew Sustain Energy Rev 45:540–555
5. Marañón E, Castrillón L, Quiroga G, Fernández-Nava Y, Gómez L, García MM (2012) Co-digestion of cattle manure with food waste and sludge to increase biogas production. Waste Manag 32:1821–1825
6. El-Mashad HM, Zhang R (101) Biogas production from co-digestion of dairy manure and food waste. Bioresour Technol 101:4021–4028
7. Qasim SR (1999) Wastewater treatment plants: planning, design and operation, 2nd edn. CRS Press LLC, London
8. Clay S, Hodgkinson A, Upton J, Green M (1996) Developing acceptable sewage screening practices. Water Sci Technol 33:229–234
9. Wid N (2012) Resource recovery from screenings through anaerobic digestion. Ph.D., School of Civil Engineering, Faculty of Engineering, University of Leeds, United Kingdom
10. Le Hyaric R, Canler J-P, Barillon B, Naquin P, Gourdon R (2010) Pilot-scale anaerobic digestion of screenings from wastewater treatment plants. Bioresour Technol 101:9006–9011
11. Horan NJ (2008) Design and operation of wastewater treatment plants for freshwater fisheries directive (FFD) compliance. Aqua Enviro Technology Transfer, UK
12. Bode H, Imhoff KR (1996) Current and planned disposal of sewage sludge and other waste products from the Ruhrverband wastewater treatment. Water Sci Technol 33:219–228
13. Huber H, Tanik AB, Gerçek M (1995) Case studies on preliminary treatment facilities at marine outfalls. Water Sci Technol 32:265–271
14. Le Hyaric R, Canler JP, Barillon B, Naquin P, Gourdon R (2009) Characterization of screenings from three municipal wastewater treatment plants in the Region Rhône-Alpes. Water Sci Technol WST 60.2:525–531
15. Le Hyaric R, Canler J-P, Barillon B, Naquin P, Gourdon R (2009) Characterization of screenings from three municipal wastewater treatment plants in the Region Rhône-Alpes. Water Sci Technol 60:525–531
16. Wid N, Horan NJ (2016) Anaerobic digestion of wastewater screenings for resource recovery and waste reduction. IOP Conference Series: Earth Environ Sci 36(1):1–7
17. Tchobanoglous G, Kreith F (2002) Handbook of solid waste management, 2nd edn. McGraw-Hill, New York
18. Mata-Alvarez J, Cecchi F, Pavan P, Llabres P (1990) The performances of digesters treating the organic fraction of municipal solid wastes differently sorted. Biol Wastes 33:181–199
19. Mata-Alvarez J, Macé S, Llabrés P (2000) Anaerobic digestion of organic solid wastes. An overview of research achievements and perspectives. Bioresour Technol 74:3–16
20. Gerardi MH (2003) The microbiological of anaerobic digesters. Wiley, Hoboken, NJ, Canada
21. Chynoweth DP, Conrad JR, Srivastava VJ, Jerger DE, Mensinger JD, Fannin KF (1985) Anaerobic processes. J (Water Pollut Control Fed) 57:533–539
22. Chynoweth DP, Turick CE, Owens JM, Jerger DE, Peck MW (1993) Biochemical methane potential of biomass and waste feedstocks. Biomass Bioenergy 5:95–111
23. Wid N, Selaman R, Jopony M (2017) Enhancing phosphorus recovery from different wastes by using anaerobic digestion technique. Adv Sci Lett 23:1437–1439
24. Cadavid-Rodriguez LS, Horan NJ (2014) Production of volatile fatty acids from wastewater screenings using a leach-bed reactor. Water Res 60:242–249
25. Gujer W, Zehnder AJB (1983) Conversion processes in anaerobic digestion. Water Sci Technol 15:127–167
26. Siegrist H, Renggli D, Gujer W (1993) Mathematical modelling of anaerobic mesophilic sewage sludge treatment. Water Sci Technol 27:25–36
27. Gavala H, Angelidaki I, Ahring B (2003) Kinetics and modeling of anaerobic digestion. Process Biomethanation 81:57–93
28. Vavilin VA, Fernandez B, Palatsi J, Flotats X (2008) Hydrolysis kinetics in anaerobic degradation of particulate organic material: an overview. Waste Manag 28:939–951
29. Cohen A, Zoetemeyer RJ, van Deursen A, van Andel JG (1979) Anaerobic digestion of glucose with separated acid production and methane formation. Water Res 13:571–580

30. Fernández B, Porrier P, Chamy R (2001) Effect of inoculum-substrate ratio on the start-up of solid waste anaerobic digesters. Water Sci Technol 44:103–108
31. Nallathambi Gunaseelan V (1997) Anaerobic digestion of biomass for methane production: a review. Biomass Bioenergy 13:83–114
32. Sharma VK, Testa C, Castelluccio G (1999) Anaerobic treatment of semi-solid organic waste. Energy Conv Manag 40:369–384
33. Ghosh S (1987) Improved sludge gasification by two-phase anaerobic digestion. J Environ Eng 113:1265–1284
34. Azbar N, Speece RE (2001) Two-phase, two-stage and single stage anaerobic process comparison. J Environ Eng 127
35. Selaman R, Wid N (2016) Anaerobic co-digestion of food waste and palm oil mill effluent for phosphorus recovery: effect of reduction of total solids, volatile solids and cations. Trans Sci Technol J 3:265–270
36. Wid N, Horan NJ (2017) Recovering resources from wastewater screenings through anaerobic digestion and phosphorus precipitation. Curr Biochem Eng 4:64–67
37. Owen WF, Stuckey DC, Healy JB, Young LY, McCarty PL (1979) Bioassay for monitoring biochemical methane potential and anaerobic toxicity. Water Res 13:485–492
38. Raposo F, De la Rubia MA, Fernández-Cegrí V, Borja R (2011) Anaerobic digestion of solid organic substrates in batch mode: an overview relating to methane yields and experimental procedures. Renew Sustain Energy Rev 16:861–877
39. Kivaisi AK, Eliapenda S (1995) Application of rumen microorganisms for enhanced anaerobic degradation of bagasse and maize bran. Biomass Bioenergy 8:45–50
40. Labatut RA, Angenent LT, Scott NR (2011) Biochemical methane potential and biodegradability of complex organic substrates. Bioresour Technol 102:2255–2264
41. Hansen TL, Schmidt JE, Angelidaki I, Marca E, Jansen JlC, Mosbæk H et al (2004) Method for determination of methane potentials of solid organic waste. Waste Manag 24:393–400
42. Viswanath P, Sumithra Devi S, Nand K (1992) Anaerobic digestion of fruit and vegetable processing wastes for biogas production. Bioresour Technol 40:43–48
43. Yang JK, Choi KM, Lee ST, Mori T (1997) Effects of hydraulic retention time on the anaerobic digestion of thickened excess sludge by sulfite in the two phase digester. Environ Eng Res 2:191–200
44. APHA (2005) Standard methods for the examination of water and wastewater. American Public Health Association, Washington, D.C.
45. Fongsatitkul P, Elefsiniotis P, Wareham DG (2012) Two-phase anaerobic digestion of the organic fraction of municipal solid waste (OFMSW): estimation of methane production. Waste Manag Res
46. Knol W, Van Der Most MM, De Waart J (1978) Biogas production by anaerobic digestion of fruit and vegetable waste. A preliminary study. J Sci Food Agric 29:822–830
47. Ghosh S, Henry MP, Christopher RW (1985) Hemicellulose conversion by anaerobic digestion. Biomass 6:257–269
48. Jerger DE, Chynoweth DP, Isaacson HR (1987) Anaerobic digestion of sorghum biomass. Biomass 14:99–113
49. Gunaseelan VN (1994) Methane production from Parthenium hysterophorus L., a terrestrial weed, in semi-continuous fermenters. Biomass Bioenerg 6:391–398
50. Babaee A, Shayegan J (2011) Anaerobic digestion of vegetable waste. Chem Eng Trans 24:1291–1296
51. Salminen EA, Rintala JA (2002) Semi-continuous anaerobic digestion of solid poultry slaughterhouse waste: effect of hydraulic retention time and loading. Water Res 36:3175–3182

Anaerobic Digestion of Food Waste

Md. Mizanur Rahman, Yeoh Shir Lee, Fadzlita Mohd Tamiri and Melvin Gan Jet Hong

Abstract The rationale of study of anaerobic digestion systems is considered, providing details of the working principles of anaerobic digestion systems for methane production as well as management of municipal solid waste, mainly kitchen waste. Background studies on the design of different types of biodigesters and theories of the production of methane gas from food waste are also discussed. The physical and chemical operating parameters for the process of methane gas production are also deliberated in this chapter since it is an essential part to be considered during design of an anaerobic biodigester. The environmental factors that have a major influence on production of methane gas from food waste and previous research work are analysed. Baseline design information is discussed to develop a suitable portable household food waste biodigester.

Keywords Anaerobic digestion · Food waste · Biomass

1 Introduction

The lifestyles in cities as well as in villages are changing rapidly due to modernization and industrial revolutions. The result is an increase in the net demand for energy. Therefore, electricity as a source of energy plays an important role in a community's daily life especially when most of the household appliances are electrical. According to a 2016 British Petroleum (BP) statistical report, primary energy consumption worldwide increased by 1.0% between the years 2014 and 2015. It is also found that about 33% of consumed energy came from fossil fuels, mainly from oil. The contribution of alternative energy was very limited at about 213 TW-h [1]. The main sources of alternative energy are wind, hydro, solar, biogas, etc. that are used globally. The energy generation or conversion cost mainly

Md.M. Rahman (✉) · Y. S. Lee · F. M. Tamiri · M. G. J. Hong
Energy Research Unit (ERU), Mechanical Engineering Programme, Faculty of Engineering, Universiti Malaysia Sabah (UMS), Kota Kinabalu, Sabah, Malaysia
e-mail: mizanur@ums.edu.my

N. Horan et al. (eds.), *Anaerobic Digestion Processes*,
Green Energy and Technology, https://doi.org/10.1007/978-981-10-8129-3_7

depends on the source of energy as well as technology. Among all the sources, the most economical abundant source is biogas since global solid waste mainly from municipalities exceeds 1.3 billion tonnes. According to World Bank 2012, on average, globally about 1.2 kg municipal solid waste is produced per capita per day, and it is projected to rise to 1.42 kg per capita per day by 2025. It is estimated that by 2025, about 2.2 billion tonnes of solid waste will need to be processed every day. This abundant source can contribute significantly to the energy sector. Poor management of municipal solid waste has a huge impact on the environment. Therefore, the effective use of these wastes may contribute a significant amount of energy generation as well as aid national development.

Malaysia is situated in the equatorial region, therefore receiving high solar radiation intensity as well as high relative humidity. This hot and humid climate contributes to the outbreak of any infectious diseases in Malaysia. As a result, the household sector needs more energy for ventilation and for cooling and Malaysia consumed a lot of electricity.

These household activities are mainly for cooling and heating and demand more energy than our ancestors would use. Hence, the consumption of energy is relatively high and might be insufficient in the near future. To overcome the demand for more energy, a sustainable solution is required within a short period of time. One of the most appropriate and valid solutions is implementation of alternative energy as replacement of non-renewable energy.

Biogas is one of the alternatives energy sources that can be used to replace non-renewable energy mainly fossil fuel. Biogas is not a single type of gas but it is a mixture of methane (CH_4), carbon dioxide (CO_2), nitrogen (N_2), oxygen (O_2), ammonia (NH_3), hydrogen (H_2), sulphur (S) compounds and carbon monoxide (CO) [2]. All kinds of biodegradable natural waste such as food waste, animal waste, forest and agriculture residues can be used to produce biogas. However, the production processes of biogas from the food wastes are different than other biomass and the conversion time also varies with types of waste [3]. The biogas production process can be speeded up by adding a suitable enzyme during the conversion process. The general biogas generation process is shown in Fig. 1.

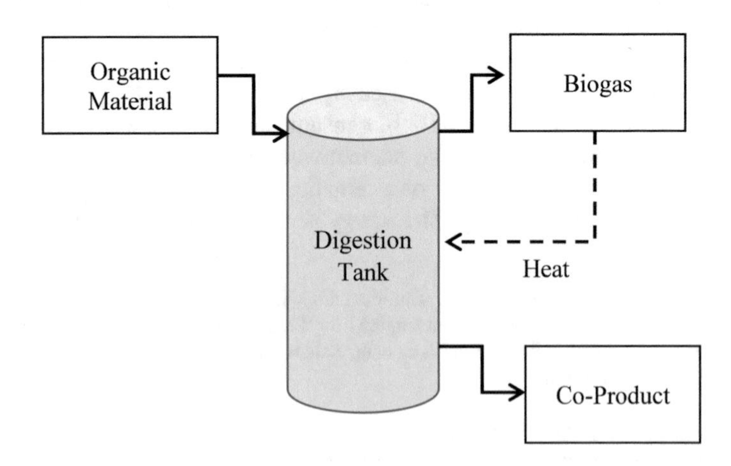

Fig. 1 Simple digester for biogas production

Biogas can be produced under anaerobic (biodegradation process in the absence of oxygen) or aerobic (biodegradation process in presence of oxygen) conditions. In the anaerobic digestion system, a series of biological processes take place in the absence of oxygen. Microorganisms in the digestion tank break down the biodegradable material and produce biogas as a product. This is a combustible gas that can be used for electricity generation as well as a source of energy for domestic and industrial use. Therefore, the anaerobic digestion system depends on environmental conditions, the operation procedure and the start-up conditions. The most challenging issues are that the bacteria must break down organic material without oxygen, with the potential of hydrogen (pH) almost neutral ($\approx$7.0); in addition, a suitable temperature and sufficient humidity, nutrients and alkalinity are needed [2]. In the anaerobic process, the organic materials are converted into biogas, which is a mixture of methane (55–75%) and carbon dioxide (25–45%) gases.

In the anaerobic process, the microorganisms break down biodegradable material through a series of biological processes before the useful end product is produced. The processes are involved hydrolysis, acidogenesis, acetogenesis and methanogenesis [4]. The pH value changes through each stage due to particular function of microorganism that desires a specific pH value. If the environment is too acidic or alkaline, the function of the microorganisms might be reduced or failed to function. It will delay the complete decomposition of the biodegradable material and reduce produced biogas. Alternatively, aerobic digestion is a biological process where bacteria in the presence of oxygen break down natural organic and biological waste, resulting in the overall reduction of natural organic material and the generation of limited quantity of stabilised cell mass as well as discharging ammonia-N and phosphate. The performance of aerobic digestion depends on solids retention time, temperature, pH value, mixing, types of solids and biodegradable material configuration.

Animal manure and human sewage sludge can be decomposed using the anaerobic digestion process. Nowadays, this process can also use to compost some municipal solid waste (MSW) as well as green waste. However, aerobic digestion can be used to process any type of organic materials. For an effective composting through aerobic digestion, the right blend of ingredients and conditions are essential, namely, that the moisture content is about 60–70% and the C/N (carbon to nitrogen) ratio is approximately 30/1. The variation in ingredients and environmental conditions might alter the performance of the aerobic digestion system. Furthermore, to ensure the presence of oxygen throughout the process a ventilation system, either passive or forced ventilation is needed to introduce air.

2 Potential of the Anaerobic Digestion Process

Malaysia is one of the fastest developing countries in Southeast Asia. Energy consumption is increasing significantly with the gross domestic product (GDP). In 2010, the GDP of Malaysia was $15,385 per capita with the economy and industry mushrooming [5]. The relationships of GDP with electricity consumption have been correlated [6]. Malaysia generates most of its electrical energy by using fossil fuel. With those inextricable relationships, it could be concluded that increases in GDP means an increase in energy demand. In correlation to that, the fluctuating constantly increasing fossil fuel price and the emission of greenhouse gases (GHG), push the development of Malaysia towards the renewable energy field. The government have started up renewable energy-powered systems and created renewable energy policies [5].

There is another knotty problem in Malaysia, which is to manage solid waste from households, which requires a well-thought-out solution. Currently, landfills are used as the main disposal place for managing those households' solid wastes, and this gradually increases every year. However, this is not a sustainable way for household waste management [7]. There is insufficient land to place those waste in near future in Malaysia [7, 8]. On average, a Malaysian creates 1.2 kg of wastes per day, of which about 74% is food waste, approximately 21% plastic and 5% other wastes [9]. Most of the solid waste produced actually comes from the kitchen, which is daily food waste. The food waste generates a high percentage of methane gas, which can be used as a source of energy [3]. In addition, the combustion of methane gas is clean and environmental friendly [10].

Methane gas could bring the benefit of a greener living environment as a useful energy source and also a sustainable way to deal with those food wastes. By contrast, it could be a reason to produce harmful greenhouse gases and may cause global warming if it is not handled properly [11]. Studies have shown that the Earth's temperature is increasing annually and the Arctic sea ice is melting due to climate changes [12]. This is a warning sign from our mother Earth. The micro level solution, using less energy at household level, is a valid and suitable option to reduce the emission of greenhouse gases as well as reduce the effect of climate change.

In contemplation of solving the problems, energy can be generated from household waste with an efficient and portable biodigester [3]. In the biodigester, the daily kitchen food waste is applied as input materials. Generally, food waste consists of cooked, uncooked and raw vegetables and fruits. The wastes are decomposed inside a confined inner compartment and generate methane gas. The gas is stored in a gas storage chamber. The end product, methane gas is required to burn with the presence of sufficient oxygen in furtherance of clean combustion [10].

This technique will help to reduce CO_2 emissions and limit the emission of one of the greenhouse gases—methane, especially from animal manure, like cow dung. In addition, during biogas production, the digested slurry is also produced as a by-product that is a micro- and macro-nutrients rich fertiliser suitable for plants.

3 Energy

In thermodynamics, energy is represented the capacity of a physical system to produce work and heat that appears in different forms. Energy cannot be destroyed but it can be transferred from one form to another form. The growth of world population and living standards in developing and developed countries are giving pressure on global energy consumption which increases on average by 2% every year [10, 13]. It is estimated that in 2025, about 8.4 billion people will live on the Earth and need a huge amount of energy [14]. This energy is harvested mainly from the renewable and non-renewable sources of energy. Unfortunately, the contribution of renewable energy is still small compared to non-renewable energy. The use of energy mainly depends on the efficiency of energy production, levels of energy conversion and the expected standard of the population [15].

In Malaysia, the growth of energy demand is very significant. In 1991, the primary energy supply was 20,611 kilotonne of oil equivalent (ktoe) and it increased to 50,658 ktoe in 2000 and 54,194 ktoe in 2013. At the same time, electricity demand also increased from 22,273 gigawatts hour (GWh) in 1991 to 60,299 GWh in 2000 and further increased to 71,159 GWh in 2003. This is mainly due to the growth of the economy and industry in Malaysia, and there is an inseparable relation between GDP with GWh [6].

3.1 Potential Sources of Energy

Malaysia is located in Southeast Asia whose energy demand increased more than 50% from 2000 to 2013. The energy demand is going to approximately double its value by 2040. Fossil fuels are the primary source of energy in Malaysia and it is predicted to increase more than 90% in 2040 [16]. In 1993, the energy source for Malaysia was only from natural gas (44.4%), crude oil (53.1%), coal and coke (0.4%) and hydropower (2.1%). Oil reserves were predicted to be finished in 15 years, if the usage of crude oil was controlled at 2.9 million barrels per day. Therefore, the Malaysia government has implemented a number of energy policies since 1974. Later in 2013, a few renewable energy sources such as biodiesel and biomass were added to the policy to strengthen it. Malaysia also has potential to use solar, municipal solid waste, small and mini-hydro, fuel cells, wind energy and geothermal power as a source of renewable energy. Palm Oil Mill Effluent (POME) can be used as the main source of biomass and biogas. It has been estimated to generate roughly 1750 MW by 2028 [5].

3.2 Biomass Sources Energy

Biomass refers to any type of animal or plant, which can be converted to energy and it can be divided into three categories: wastes, standing forests and energy crops [17]. Biomass energy resources are produced in areas where plantations are grown for fuel or heating purposes [18]. The biomass resources and its potential for producing bioenergy is still not yet fully utilised today [19]. Biomass could be the most powerful and ideal energy source in the near future if it is fully utilised. Properly managed it minimises harm to the environment. Biomass is categorised as an energy source that does not emit carbon dioxide (CO_2) to the environment when decomposed. The carbon cycle is a complex global process when biomass is breaking down and the organic carbon is recycled. Through this, the CO_2 produced is able to be absorbed by the plants during growth [14].

Energy can be harvested from biomass resources either by mechanical or natural processes. These techniques are used to convert biomass resources into liquid or gaseous fuel [19]. The converted biomass fuel can be used efficiently to run vehicles or generate electricity as a replacement for fossil fuel. Figure 2 summarises the forms of biomass conversion pathways.

3.2.1 Mechanical Process

Most of the biomass is required to be preprocessed such as compression, chopping, preheating, etc. before it can be put it into a conversion process [14]. Thermochemical processes such as combustion, gasification and pyrolysis are used to convert biomass into fuel. Thermal conversion takes place effectively if the moisture content in biomass is less than 15%. Combustion of biomass is a process to produce steam and power. Direct combustion of biomass is efficient if the moisture content is more than 15%. Gasification is one of the most suitable techniques to ensure efficient biomass combustion. A gasifier is used to convert biomass into producer gas that contains mainly carbon monoxide, hydrogen, carbon dioxide,

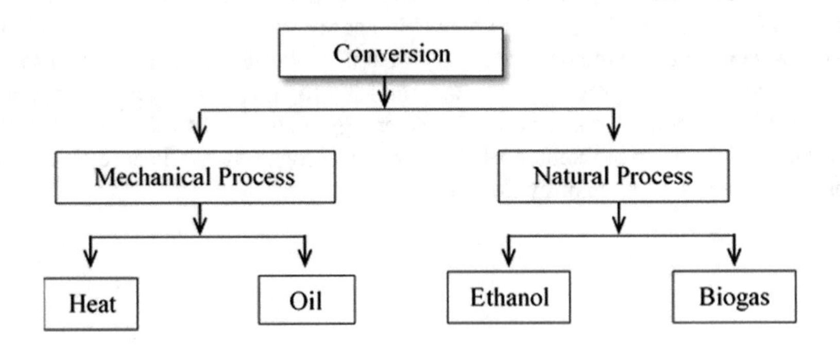

Fig. 2 Biomass conversion pathways

methane and nitrogen. A gasifier is a chemical reactor where biomass goes through several complex physical processes and chemical reactions take place under an incomplete supply of air. As a result, producer gases are generated which can be used for heating water to produce steam and run a power plant to generate electricity. The producer gas is not a single type of gas it is a mixture of combustible and non-combustible gases. The volume of combustible and non-combustible gases depends on the types of biomass that is used a raw material. The heating value of the producer gas varied from 4.5 to 6 MJ/m^3, it also depends on the biomass material. Therefore, the gasifier, gas flow direction and capacity are considered. Gasifiers are classified into three groups: updraft, downdraft and cross-draft.

During the gasification process, biomass goes through drying, oxidation, pyrolysis and mass reduction processes. Among the processes, pyrolysis is known as a thermochemical process, which helps decomposes biomass material. Biomass materials can be converted into a liquid that can be applied in transportation or as a type of chemical feedstock [10]. Furthermore, biomass waste such rice husk, sawdust and leaves have a high bulk volume. Hence, it is very hard to transport from one place to another place. In addition, the heating values of loose biomass per unit volume are very low compare to other fuels. The loose biomass can be compressed and increased the heating value per unit volume which is known briquetting [19].

3.2.2 Natural Process

This process is also known as a biological process, which involves a series of biochemical reactions in the presence of enzymes or enzymatic reactions [10]. The biochemical process is very sensitive and perceptive towards the surrounding environment. This process is particularly depended on the presence of enzymes, microorganisms or bacteria, which are used for breaking down the complex substances into their simplest form. Enzymes act like key for a padlock and only the right key can be used to open the specialised padlock. There are two main types of biochemical processes that can be used to turn biomass resources into synthetic fuels. The biochemical processes are known as fermentation and anaerobic digestion [18]. The differences between them are the involvement of oxygen throughout the process. The operational parameters are different according to the specific process on the grounds of how the enzymes work [21].

3.3 Biomass and Biogas

Many developing and underdeveloped countries in the world, mainly in rural areas, use biomass as a source of energy [21]. The efficiency of direct combustion of biomass varies from 10 to 15% depending on the moisture content in the biomass. The combustion efficiency can be increased up to 30% by using improved stoves. However, the direct combustion of biomass generates harmful carbon monoxide

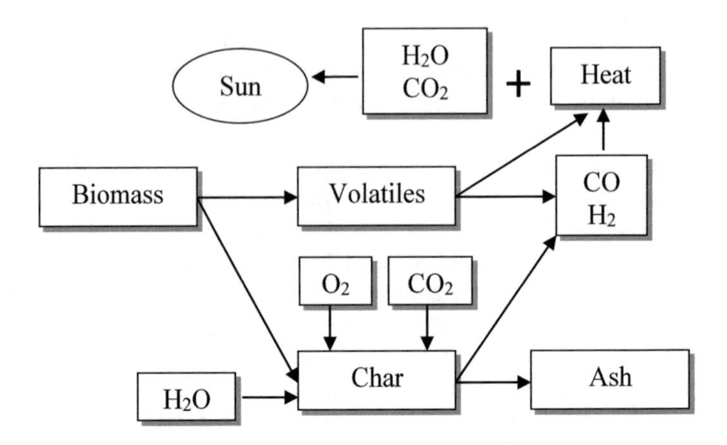

Fig. 3 Combustion of biomass and the carbon cycle

(CO) and hydrogen (H_2) gas (Fig. 3). Biomass can be used to produce biogas. In the biogas production process, biodegradable raw materials are decomposed either in the presence or absence of oxygen. The produced biogas can be used as a source of energy. The process is more commonly named as aerobic or anaerobic digestion, respectively.

3.3.1 Anaerobic Digestion

During anaerobic digestion, biogas is produced that is an end product from biomass raw material which has a high volume of carbon compound [22]. Food wastes, palm oil wastes, animal manures, aquatic and woods can be used as input materials for the anaerobic digestion process [23]. The biogas that is produced during the digestion processes is predominantly a mixture of methane and carbon dioxide with a negligible amount of nitrogen, hydrogen, hydrogen sulphide, ammonia and oxygen [10]. The performance of the biodigester can be determined from the amount of collected gases. Studies showed that the biogas is only flammable when methane gas is >45% in the producer gas [24]. Methane is a colourless, odourless gas and gives a blue flame during burning. The combustion of biogas, mainly the burning of methane gas, with sufficient presence of oxygen becomes water (H_2O) and carbon dioxide (CO_2) [10, 22]:

$$CH_4 + 2O_2 \rightarrow CO_2 + 2H_2O + 192\ \text{kcal mol}$$

The combustion of biogas is clean; therefore, it is known as an environmentally friendly source of energy and can be used as a fuel for electricity generation plants. Therefore, the effective use of biogas can contribute to national development.

3.3.2 Solid Waste Digestion Processes

Anaerobic digestion is a suitable option to decompose biodegradable waste, crop waste and wastewater to generate energy. It is a decomposition process under controlled conditions and in the absence of oxygen. Mesophilic or thermophilic temperatures are suitable for anaerobic digestion and facultative anaerobic bacteria and archaea species occur naturally, which able to convert the wastes to biogas [25, 26].

Anaerobic digestion is a microbial process of decomposition of organic matter, in sealed reactors tanks, that reduces the percentages of the organic waste and creates biogas and fertiliser [27]. The advantages of using anaerobic digestion over aerobic digestion are that anaerobic decomposition requires a lower energy and operation cost. On the other hand, anaerobic digestion needs a long period of time; thus, a larger hydraulic retention time is demanded [28].

3.4 Principles of Anaerobic Digestion

There are four different steps during anaerobic digestion for the deterioration of carbon-based raw material which are hydrolysis, acidogenesis, acetogenesis and methanogenesis [25]. The stages of the anaerobic process are shown in Fig. 4.

Hydrolysis

In this process, with the help of exoenzymes and bacterial cellulosomes, the complex carbohydrates, fats and proteins are hydrolysed to their monomeric forms

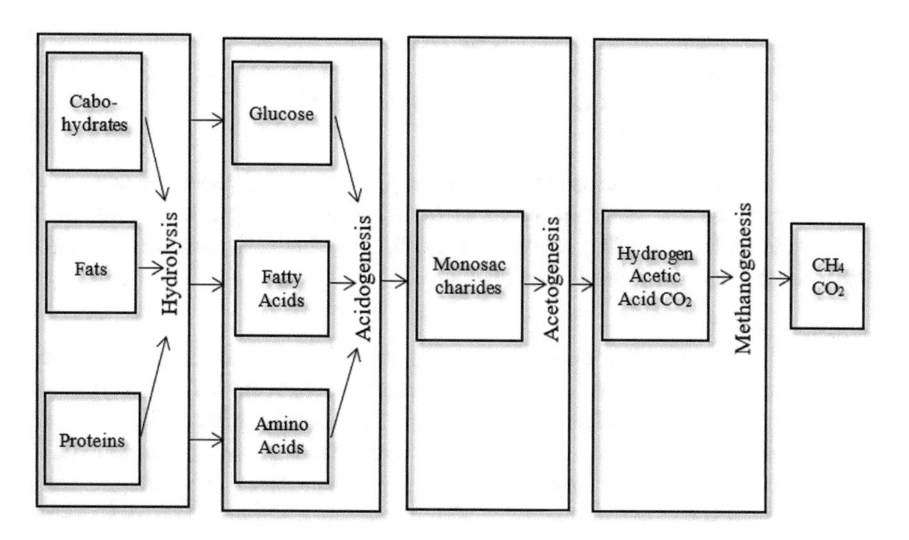

Fig. 4 Stage of anaerobic digestion

[21]. The enzymes act as a catalyst to speed up the digestion process, where carbohydrates become simple glucose, proteins hydrolyse to amino acids and lipids to long-chain fatty acids [29, 30].

Acidogenesis

This step is also known as fermentation since glucose, amino acids, fatty acids and peptides are fermented to volatile fatty acids by anaerobic bacteria [10]. Volatile fatty acids such as acetic acid, propionic acid, butyric acid, isobutyric acid, valeric acid, isovaleric acid, caprionic acid, alcohols, hydrogen and carbon dioxide are produced through this process [21]. The acidogens reproduce at a rapid rate which causes the pH to drop which is harmful if the volatile acids produced are not removed by the next stage [31].

Acetogenesis

The higher volatile fatty acids and other organic compounds are transformed to acetic acid, hydrogen and carbon dioxide by the acetogenic bacteria [31].

Methanogenesis

Methanogens convert the intermediate products into methane and carbon dioxide [21]. There are two types of methanogens process in this stage: one type is acetoclastic methanogens that nurture along with acetic acid and generate about 70% of methane gas, and the other type is hydrogen exploiting methanogens that reduced carbon dioxide and hydrogen [31].

3.4.1 Operational Conditions in Anaerobic Digestion

Methane production from biomass involves varies types of microorganisms. The performance of the anaerobic digestion processes depends mainly on the volume of biomass, temperature, mixing, carbon–nitrogen ratio, hydraulic retention time and the potential of hydrogen (pH). These parameters are also important to maintain the balance of microorganisms in the digestion system [21].

Volume

The total reactor volume used for batch tests is inversely related to the number of replicate samples that could be tested at the same time using a fixed amount of sludge and substrate. The useful reactor for biochemical methane potential (BMP) evaluation was always less than 1L [23].

Temperature

There is a wide range of temperature where anaerobic digestion can occur, which are classified into three groups, namely, as psychrophilic (below 20 °C), mesophilic (optimum 30–40 °C) and thermophilic (above 50 °C) [31]. Mesophilic and thermophilic are more favoured compared to psychrophilic because the reaction rate is higher at both temperature ranges [31]. Mesophilic anaerobic processes are common due to the stable reaction process, high in microbial activities and easy to start up [10].

Mixing Substrate

It is important to retain consistent substrate concentration, temperature and other environmental factors to avert scum formation and solids deposition. Mixing through gas circulation and mechanical stirrers are depending on the final total solids concentrations of the systems [32].

Carbon–Nitrogen Ratio (C/N Ratio)

The performance of the anaerobic digestion process relies on the concentrations of carbon and nitrogen, where carbon is the energy source for the microorganisms and nitrogen enhances microbial growth. It is generally lost as ammonia gas when the amount of nitrogen exceeds the microbial requirement. For the ideal condition, the ratio of carbon to nitrogen is about 30:1 in the raw material [32].

Hydraulic Retention Time (HRT)

Hydraulic retention time is defined as the period the raw material remains in the digester and is a key design parameter used that depends on the chemical oxygen demand (COD) and biochemical oxygen demand (BOD) of the feedstock and slurry. The optimal hydraulic retention time for full biological conversion using mesophilic digestion is 15–30 days [33].

$$\text{Hydraulic Retention Time} = V_r/Q$$

where

V_r Volume of reactor (m^3),
Q Influent flow rate.

The Potential of Hydrogen (pH) Value

The pH value of the digester environment is a very important factor and plays an important role in biomass digestion and methane gas production. Bacteria are very sensitive towards acidic conditions. The pH value for each stage is different due to the volatile fatty acids, carbon dioxide and small chain carbonates; therefore, a buffer solution such as lime may be added to stabilise the pH at a neutral value [33].

3.5 Anaerobic Biodigester

Biodigesters such as fixed dome, floating drum and plug flow are used to produce methane gas and usually are of large size. There are a few types of small-scale biodigesters suitable for households' waste [21]. The biodigester is a waterproof, sealed chamber usually cubical or cylindrical and built using brick, concrete, steel or plastic, to allow the process of transforming organic waste into biogas to take place [24].

3.5.1 Fixed Dome Digester

A fixed dome digester is known as 'Chinese' or 'hydraulic' digester. In this digester, biomass raw material is fed through the inlet pipe, the biogas produced accumulates at the upper part of the digester and the gas pressure formed compresses the slurry to the outlet pipe [21, 34]. Figure 5 shows the fixed dome digester and its different parts. There are two different types of fixed dome digester models; Janata model and Deenbandhu model are commonly used. The main difference between Janata model and Deenbandhu model is the shape of the floor bottom. The digestion tank of Deenbandhu model is designed as a hemispherical shape, while Janata model has a flat bottom. The disadvantage of these biodigesters is once installed, it is hard to move or replace them [21]. In addition, the digester tank leaks easily when a high gas pressure develops in the biogas chamber [35, 36].

The advantages of fixed dome digesters are lower construction cost, they do not contain moving and rust-prone steel parts, longer lifetime, less temperature variation and more space-saving [35, 36].

3.5.2 Floating Drum Digester

This type of digester is designed with a movable inverted drum, which is known as a gas collector and able to move up and down depending on the gas produced at the

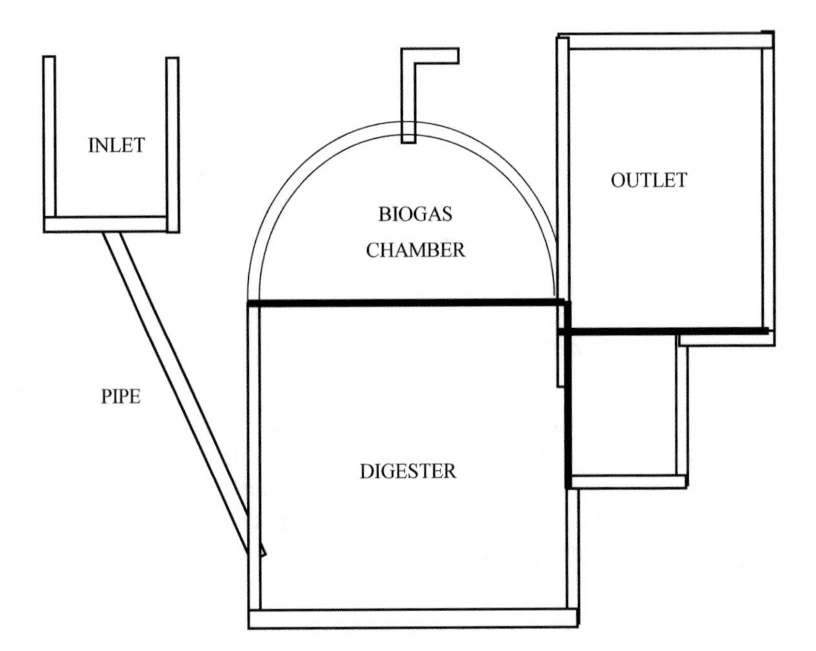

Fig. 5 Fixed dome anaerobic digester

top of the digester tank (Fig. 6). The gas pressure can be increased by adding load to the inverted drum that is needed for the gas flow through the piping, for further usage. In this digester, the amount of gas accumulated can be predicted from the position of the drum. The average size of the digester tank is around 1.2 m^3. The difference between a floating drum digester and fixed dome digester is the gas collection part. The floating drum digester uses a floating drum, steel, gas collector; however, the fixed dome digester uses a fixed steel gas collector. The drum can rust over time and reduce the movement of the drum as well as reduce the performance of the digester [21, 37].

3.5.3 Plug Flow Digester

This type of digester is designed as a long and narrow horizontal tank in which biomass is constantly added at the influent. Continuous digestion takes place in the plug flow digester; therefore, the production of gas is continuous. Usually, the plug flow digester vessel is five times longer than it is wide. It is made of concrete, steel or fibreglass [38]. The inlet and outlet of the digester are put at two different ends and placed above the ground; the other parts of the digester are buried inclined in the ground as shown in Fig. 7. The main advantage of plug flow is simple, easy to operate and economical to instal [38].

According to Cheng et al. [28] and Deng et al. [35], the fixed dome, floating drum and plug flow digesters can be used for food waste. All the digesters have advantages and disadvantage as presented in Table 1.

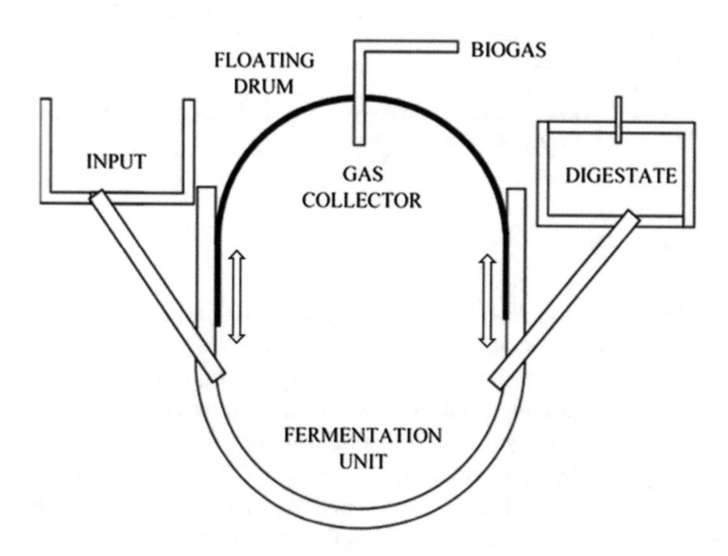

Fig. 6 Floating drum digester

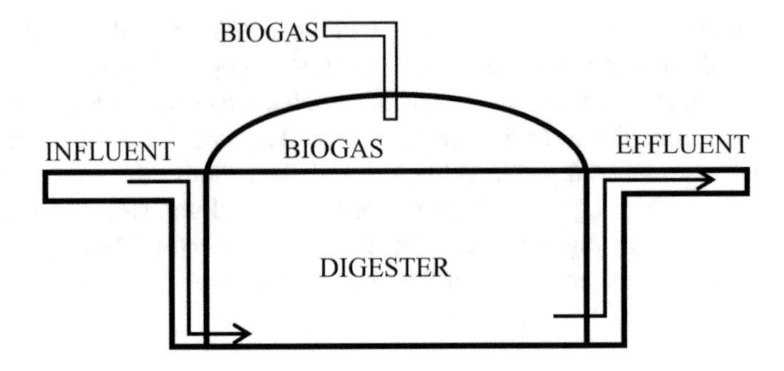

Fig. 7 Plug flow anaerobic digester

Table 1 Advantage and disadvantage of fixed dome, floating drum and plug flow biodigesters

Digester type	Advantages	Disadvantages
Fixed dome	• Low starting cost • Long lifetime • Very less rusting parts • Space-saving • Low maintenance required during operation • Small area required if built underground	• High technical skills needed to construct gas-tight condition • Hard to repair if leak • Heavy construction materials demanded • Quantity of gas produced not easily detected • High gas pressure
Floating drum	• Easy operation and construction • Gas volume noticeable • Can maintain fixed gas pressure	• Expensive due to steel drum • Corrosion of steel drum caused shorten lifetime • Required high maintenance cost for painting the steel drum
Plug flow	• Low cost and construction • Transportable • Easy maintenance • More independent towards climate change if compare to fixed dome type	• Short lifetime • Easily damaged • Low gas pressure • Less eco-friendly

3.6 *Food Waste*

The bio-methanisation of food waste is divided into four main categories based on the food waste main components such as proteins, lipids, carbohydrates and celluloses. The rate of production of methane is highest from food wastes that have excess lipids but the reaction needs a longer time. Food waste with proteins takes a little longer, then celluloses and lastly carbohydrates [33]. Food waste has a high content of organic matter, comprising crude lipid (22.8–31.45%), crude protein (14.71–28.64%) and carbohydrates on a dry matter basis [39]. Household food wastes categories are rich in proteins and lipids from meat, egg, cheese and fish or high in carbohydrates from bread, flour and potatoes. Vegetables and fresh fruits are usually rich in carbohydrates except for mushroom, spinach and dry fruits which are full of proteins [33].

3.7 *Biodigester for Food Waste*

According to Deng et al. [35] and Tapase et al. [40], the fixed dome biodigester can be used successfully under mesophilic condition for food waste digestion. A floating drum anaerobic biodigester under mesophilic condition can also be used for food waste [41]. To produce methane effectively, Asmare [42] combined food waste with human waste before putting it into the fixed dome digester tank under mesophilic conditions. Lama et al. [43] pretreated the floating drum biodigester with cow dung before placing food waste in it. This study showed that the numbers of microorganisms were increased significantly after pretreatment. The anaerobic digestion of the food waste at mesophilic temperature was faster and a specific amount of methane gas is produced under this process [43]. Ogur and Mbatia [3] selected only uncooked kitchen waste as their organic matter to put into the biodigester and operated it under mesophilic conditions. The experiment accumulated 24 m^3/day biogas for an input of waste of approximately 100 kg. Vij [44] chose cooked rice, vegetables and vegetable waste as the source of organic matter and managed to produce methane gas in only 52 h. The summary of the outcome of these experiments is presented in Table 2.

Table 2 Summary of food waste biodigester

Biodigester type	Food waste type	References
Fixed dome	Food waste	[35]
Fixed dome	Food waste	[40]
Fixed dome	Food and human waste	[42]
Fixed dome	Uncooked kitchen waste	[3]
Fixed dome	Cooked rice and vegetables, and vegetable wastes	[44]
Floating drum	Food waste (pretreatment with cow dung)	[43]
Floating drum	Food waste	[41]

3.8 Conclusion and Future Remarks

The huge amounts of food waste and other biodegradable biomass are responsible for severe environmental pollution in many countries in the world. Therefore, food waste management is becoming a major issue for every city as well as for society. Nowadays, it is of serious economic and environmental concern; hence, many organisations are working for effective solution for food waste management. Many researchers all over the world are putting their efforts into finding a sustainable solution and conversion technologies to manage this waste. The main components of food wastes are proteins, lipids, carbohydrates and celluloses. There are many different ways food waste can be managed. Among all the management methods, the anaerobic digestion process is one of the valid options that can manage food waste effectively as well as generate a by-product, biogas that can be used as a source of energy. In the anaerobic digestion process, the microorganisms effectively break down the natural biomass materials and food waste in the absence of the oxygen and produce biogas, mainly methane. The methane gas is an appropriate solution for rural household energy supply, mainly for cooking and heating. This methane gas can also be used as a source fuel for generators to generate electricity or as a vehicular fuel but the biogas needs to fulfil certain condition of combustion. The performance of the anaerobic digestion can be improved significantly by reviewing several physical, thermochemical, biological or combined pretreatments. The food waste digestion performance also depends on the rate of reaction in the digestion chamber and hydrolysis process. Food waste with others biomass significantly enhances the production of biogas, mainly methane. In the anaerobic digestion system, the percentage of combustible gas mainly methane and non-combustible gas mainly CO_2 are about 69 and 29%, respectively [45].

Additionally, there are a few parameters that need to be considered to improve the performance of the anaerobic digestion process. First, a large volume of digester tank should be used with centralised gas storage. This will increase the methane gas concentration at the outlet. Furthermore, suitable enzymes can be used to enhance the reaction rate for the digestion of food waste. Different combinations of food waste and biomass material have a significant effect on the digestion rate. Therefore, the enzyme can be suitable for all kind of food wastes, which has altered C–H–O–N ratio.

References

1. Statistical Review of World Energy (SRWE) (2016) BP Statistical review of world energy 2016, 65th edn, vol 5
2. Schirmer WN, Jucá JFT, Schuler ARP, Holanda S, Jesus LL (2014) Methane production in anaerobic digestion of organic waste from recife (Brazil) landfill: evaluation in refuse of different ages. Braz J Chem Eng 31(2):373–384

3. Ogur EO, Mbatia S (2013) Conversion of kitchen waste into biogas. Int J Eng Sci (IJES) 11 (2):70–76
4. Vinnerås B, Schönning C (2011) Microbial risks associated with biogas and biodigestor sludge. In: Encyclopedia of environmental health, pp 757–764
5. Mekhilef S, Barimani M, Safari A, Salam Z (2014) Malaysia's renewable energy policies and programs with green aspects. Renew Sustain Energy Rev 40:497–504
6. Jafar AH, Al-Amin AQ, Siwar C (2008) Environmental impact of alternative fuel mix in electricity generation in Malaysia. Renew Energy 33(10):2229–2235
7. Moh Y, Abd Manaf L (2017) Solid waste management transformation and future challenges of source separation and recycling practice in Malaysia. Resour Conserv Recycl 116:1–14
8. Fauziah SH, Agamuthu P, Municipal solid waste management in Malaysia: strategies in reducing the dependency on landfills http://repository.um.edu.my/27036/1/Banda%20Aceh%20Fullpaper.pdf
9. Budhiarta I, Siwar C, Basri H (2012) Advanced science information technology current status of municipal solid waste generation in Malaysia. Int J Adv Sci Eng Inf Technol 2(2):16–21
10. Cheng J (ed) (2010) Biomass to renewable energy processes. Taylor & Francis Group, Raleigh, North Carolina
11. Prasad PVV, Thomas JMG, Narayanan S (2017) Global warming effects. In: Encyclopedia of applied plant sciences, pp 289–299
12. Matishov GG, Dzhenyuk SL, Moiseev DV, Zhichkin AP (2016) Trends in hydrological and ice conditions in the large marine ecosystems of the Russian Arctic during periods of climate change. Environ Dev 17:33–45
13. Ngô C, Natowitz J (2016) Our energy future: resources, alternatives and the environment. Wiley
14. Golušin M, Dodić S, Popov S (2013) Sustainable energy management. Academic Press
15. Kreith F, Kreider JF, Krumdieck S (2010) Principles of sustainable energy: mechanical and aerospace engineering series. CRC Press
16. Birol F, Bromhead A, Blasi A, Ikeyama S (2015) Southeast Asia energy outlook—world energy outlook special report 2015. www.iea.org/publications/freepublications/publication/WEO2015_SouthEastAsia.pdf
17. Schobert H (2014) Energy: the basics. Routledge
18. Cassedy ES, Grossman PZ (1998) Introduction to energy: resources, technology and society, 2nd edn. Cambridge University Press, Cambridge, UK
19. Letcher TM (2008) Future energy: improved, sustainable and clean options for our planet. Elsevier Science
21. Rajendran K, Aslanzadeh S, Taherzadeh MJ (2012) Household biogas digesters—a review. Energies 5(8):2911–2942
22. Ingole NW (2015) Development and design of biogas plant for treatment of kitchen waste. Int J Res Eng Sci Technol (IJRESTs)—Civ Eng 1:1–5
23. Raposo F, De La Rubia MA, Fernández-Cegrí V, Borja R (2012) Anaerobic digestion of solid organic substrates in batch mode: an overview relating to methane yields and experimental procedures. Renew Sustain Energy Rev 16(1):861–877
24. Ghosh R, Bhattacherjee S (2013) A review study on anaerobic digesters with an Insight to biogas production 2(3):8–17
25. Bensah EC, Ahiekpor J, Antwi E, Ribeiro J (2015) A study of the effluent quality of excrement-based biogas plants in Ghana. Int J Sci Eng Appl Sci (IJSEAS) 1(6)
26. Suanon F, Sun Q, Li M, Cai X, Zhang Y, Yan Y, Yu C-P (2017) Application of nanoscale zero valent iron and iron powder during sludge anaerobic digestion: impact on methane yield and pharmaceutical and personal care products degradation. J Hazard Mater 321:47–53
27. Jyothilakshmi R, Prakash SV (2016) Design, fabrication and experimentation of a small scale anaerobic biodigester for domestic biodegradable solid waste with energy recovery and sizing calculations. Proc Environ Sci 35:749–755

28. Cheng S, Li Z, Mang HP, Huba EM, Gao R, Wang X (2014) Development and application of prefabricated biogas digesters in developing countries. Renew Sustain Energy Rev 34: 387–400
29. Sørensen B (1991) A history of renewable energy technology. Energy Policy 19(1):8–12
30. Twidell J, Weir T (2015) Renewable energy resources. Routledge
31. Minteer SD (2011) Biochemical production of other bioalcohols: biomethanol, biopropanol, bioglycerol, and bioethylene glycol. In: Handbook of biofuels production, pp 258–265
32. Hilkiah Igoni A, Ayotamuno MJ, Eze CL, Ogaji SOT, Probert SD (2008) Designs of anaerobic digesters for producing biogas from municipal solid-waste. Appl Energy 85 (6):430–438
33. Mir MA, Hussain A, Verma C (2016) Design considerations and operational performance of anaerobic digester: a review. Cogent Eng 28(1):1–20
34. Kaur H, Sohpal VK, Kumar S (2017) Designing of small scale fixed dome biogas digester for paddy straw. 7(1)
35. Deng L, Liu Y, Zheng D, Wang L, Pu X, Song L, Long Y (2016) Application and development of biogas technology for the treatment of waste in China. Renew Sustain Energy Rev 70(2017):845–851
36. Kalia AK, Kanwar SS (1998) Long-term evaluation of a fixed dome Janata biogas plant in hilly conditions. Bioresour Technol 65(1):61–63
37. Kalia AK, Singh SP (1999) Case study of 85 m^3 floating drum biogas plant under hilly conditions. Energy Convers Manage 40(7):693–702
38. Ramatsa IM, Akinlabi ET, Madyira DM, Huberts R (2014) Design of the bio-digester for biogas production : a review, vol II, pp 22–24
39. Meng Y, Luan F, Yuan H, Chen X, Li X (2017) Enhancing anaerobic digestion performance of crude lipid in food waste by enzymatic pre-treatment. Bioresour Technol 224:48–55
40. Tapase R, Phutane S, Pawar P, Sonawane P, Chavan VM (2016) Design of fixed dome domestic bio digester for degradation of kitchen waste using mesophilic & thermophilic reactions (anaerobic). Int J Mech Eng Technol (IJMET) 7(2):52–58
41. Voegeli Y, Lohri C, Kassenga G, Baier U, Zurbrügg C (2009) Technical and biological performance of the Arti compact biogas plant for kitchen waste-case study from Tanzania. In: Margherita di Pula S (ed) 12th international waste management and landfill symposium. Cagliari, Italy, pp 5–9
42. Asmare M (2014) Design of cylindrical fixed dome bio digester in the condominium houses for cooking purpose at Dibiza site East Gojjam, Ethiopia. Am J Energy Eng 2(1):19–22
43. Lama L, Lohani SP, Lama R, Adhikari JR (2012) Production of biogas from kitchen waste, vol 2, pp 14–18
44. Vij S (2011) Biogas production from kitchen waste. In: A seminar report submitted in partial fulfillment of the requirements for Bachelor of Technology (Biotechnology), unpublished. Seminar report, Department of Biotechnology and Medical Engineering, National Institute of Technology, Rourkela, July
45. Cheng J (ed) (2009) Biomass to renewable energy processes. CRC press

Overview of Biologically Digested Leachate Treatment Using Adsorption

I. Azreen and A. Y. Zahrim

Abstract Biological process is effective in treating most biodegradable organic matter present in leachate; however, a significant amount of ammonia, metals and refractory organic compounds may still remain in this biologically digested leachate. This effluent cannot be released to receiving bodies until the discharge limit is met. Several physical/chemical processes have been practiced as post-treatment to remove the remaining pollutants including coagulation–flocculation, oxidation and adsorption. Adsorption is often applied in leachate treatment as it enhances removal of refractory organic compounds. This chapter will focus on works related to adsorption as one of the commonly used methods to treat biologically digested leachate further down to acceptable discharge limit.

Keywords Leachate · Landfill · Adsorption · Biologically digested

1 Introduction

Leachate is defined as a liquid generated mainly by the percolation of precipitation water passing through the cap of the completed landfill, or through the open landfill site. To be more specific, it is a soluble mineral and organic compound that is formed as water penetrates into the waste layers, extracting along contaminants as it flows and initiates a mass transfer through complex interplay between the biogeochemical and hydrological reactions for producing high moisture content that could activate the liquid flow [1].

Leachates usually contain massive amounts of organic contaminants, which can be measured as chemical oxygen demand (COD) and biochemical oxygen demand (BOD), suspended solid, halogenated hydrocarbons, ammoniacal nitrogen, inorganic salts and various type of heavy metals. These contaminants are of major

I. Azreen · A. Y. Zahrim (✉)
Chemical Engineering Programme, Faculty of Engineering, Universiti Malaysia Sabah, Jalan UMS, 88400 Kota Kinabalu, Sabah, Malaysia
e-mail: zahrim@ums.edu.my

N. Horan et al. (eds.), *Anaerobic Digestion Processes*,
Green Energy and Technology, https://doi.org/10.1007/978-981-10-8129-3_8

Table 1 Acceptable discharge limits for leachate in several countries [3–5]

Parameters	Discharge limit (mg/L)						
	Malaysia	USA	Hong Kong	France	Germany	China	South Korea
COD	400	NA	200	120	200	100	50
BOD_5	20	NA	800	30	20	30	NA
NH_3–N	5	NA	5	NA	NA	25	50
Total phosphorus	NA	NA	25	NA	3	3	NA
Total nitrogen	NA	NA	100	30	70	40	150
Cd(II)	0.01	0.01	0.1	NA	0.1	0.01	NA
Cr(III)	0.2	NA	0.1	NA	0.5	–	NA
Cr(VI)	0.05	0.05	NA	NA	0.1	0.05	NA
Cu(II)	0.2	0.07	1.0	NA	0.5	–	NA
Zn(II)	2	0.3	0.6	NA	2	–	NA
Ni(II)	0.2	0.013	0.6	NA	1.0	–	NA
Pb(II)	0.1	0.03	NA	NA	0.5	0.1	NA
Ag(II)	0.1	0.05	0.6	NA	NA	–	NA

concern to be treated as they are listed as the parameters of leachate discharge limit to be met by landfill operators. Table 1 shows the increasingly strict discharge standards in several countries. If not properly treated and safely disposed of, landfill leachate could cause a significant threat to water bodies as it can contaminate the surface- and groundwater and pollute the receiving waters [2].

The amount of the rainfall infiltrating to the waste through the cover, the absorptive capacity of the waste, the weight of absorptive waste and any removal of the leakage via seepage or discharge are among factors to be considered in estimating the generation rate of leachate [6]. Leachate composition depends on certain factors such as design and operation of landfill, waste composition [7], the availability of moisture and oxygen, landfill age and site hydrology [1]. Information of leachate quality is also useful in designing appropriate liner system in sanitary landfill and leachate treatment. Rafizul and Alamgir [8] characterize the leachate and investigate the influence of tropical climatic effects on leachate characteristics of lysimeter studies at various seasons. The study demonstrates that leachate quality is affected by lysimeter operational mode.

Table 2 represents classification of landfill leachate according to the composition changes. Young landfill leachate (age <5 years) is generally characterized by high BOD and COD concentrations, high ratio of BOD/COD, moderately high ammonium nitrogen and a pH value as low as 4. Stabilized or old leachate (age >10 years) is presented by a high refractory compounds which are not easily degradable, such as humic substances and fulvic-like fractions, moderately high strength of COD, high strength of ammonia nitrogen and a low BOD/COD ratio of less than 0.1 [9, 10, 12]. Quite differently, Alvarez et al. [11] classify young leachate as less than 1 year old with BOD/COD ratio ranging from 0.5 to 1.0,

Table 2 Classification of landfill leachate by age [9, 10, 12]

Type of leachate	Young	Intermediate	Stabilized
Age (years)	<5	5–10	>10
pH	<6.5	6.5–7.5	>7.5
COD (mg/L)	>10,000	4000–10,000	<4000
BOD_5/COD	0.5–1.0	0.1–0.5	<0.1
Ammonia nitrogen (mg/L)	<400	NA	>400

and old leachate as more than 5 years old with BOD/COD ratio ranging from 0 to 0.3. In this paper, biologically digested leachate refers to young or intermediate leachate that has purposely undergone biological treatment or stabilized leachate. Over the time, the microorganisms break down the organic matters in leachate into CH_4 and CO_2. Hence, the organic compounds become less biodegradable, and the leachate becomes stabilized. Like stabilized leachate, biologically treated leachate is also lack of biodegradable materials.

From Table 2, it is noticeable that ammonia nitrogen in stabilized leachate is at a high concentration and usually requires post-treatment, whereas COD concentration, biodegradability and heavy metals have been reduced in stabilized leachate.

2 Treatment of Leachate

Inappropriate landfill management can lead to serious environmental problems through release of leachate. Landfill leachate is among the most challenging effluents to treat owing to its complex composition, highly variable characteristics and strength [13, 14]. Most researchers discuss leachate treatment technologies by classifying them into two basic types: biological and physical/chemical [3, 15, 16].

Both physical/chemical and biological treatment processes have their advantages and disadvantages; therefore, typically these processes are combined for a more effective leachate treatment [14, 17, 18]. Selection of an appropriate and adequate leachate treatment process is dependent on contaminants that need to be eliminated from the leachates. According to Keenan et al. [19], usage of biological treatment alone is difficult because of the leachate characteristics. Low removal efficiency of

Table 3 Typical leachate characteristics from landfills in different countries

Country	COD (mg/L)	BOD_5 (mg/L)	BOD_5/COD	pH	NH_3–N (mg/L)	References
Belgium	706–1846	20–50	0.02–0.03	8.0–8.5	–	[21]
China	2817	150	–	8.6	2000	[2]
Malaysia	2740	193.2	0.07	8.3	1113.2	[22]
France	550	–	0.03	7.5	390	[23]

each biological treatment system is due to substantial presence of hard to remove high-molecular-weight organics and presence of organics, inorganic salts and metals which inhibit the activated sludge microorganisms. Therefore, landfill leachate characteristics must be known to understand the inconstant performance in leachate treatment either by biological or physical/chemical process [20]. Table 3 shows typical leachate characteristics from landfills in different countries.

Apart from undergoing physical/chemical and biological treatment, leachate can also be treated by recycling or combined treatment with municipal wastewater [3, 15]. As an option, leachate can also be used as fertilizer for wheat plant (*Triticum aestivum*) [24]. Table 4 summarizes types of landfill leachate treatments.

Selection of treatment technologies to be applied in treating leachate is depending on the type of contaminants to be treated. In a study by Bashir et al. [25], the efficiency of a treatment process is determined based on colour, COD and NH_3–N percentage removals. The best removal of colour, COD and NH_3–N from leachate which are 96.8, 87.9 and 93.8%, respectively, is achieved using ion exchange treatment via cationic/anionic sequence as compared with other treatment methods. Coagulation–flocculation and AOPs were effective for COD and colour removal from leachate but not for NH_3–N. Biological treatment manages to remove 71% NH_3–N, but only removes 29% COD and 22% colour. Better removal of both NH_3–N (92.6%) and COD (68.4%) from stabilized leachate is achieved using adsorption via a new carbon–mineral composite.

It is clear from this finding that different technologies provide different efficiencies for contaminant removal. Moreover, due to the inconsistency in leachate composition as indicated as parameters in Table 1, no single effective treatment is available for all landfill leachate. Risks associated with the leachates dispersion are caused by high pollutant concentrations and the non-homogenous nature of the substances [16].

Table 4 Types of landfill leachate treatments commonly used

Recycling [3, 15]	
Combined treatment with municipal wastewater [3, 15]	
Biological treatment	Aerobic treatment [3, 15, 16, 20] Anaerobic treatment [3, 15, 16, 20]
Physical/Chemical treatment	Dissolved air flotation [3] Air stripping [3, 4, 15] Coagulation–flocculation [3, 4, 15, 20] Chemical precipitation [3, 4, 15] Chemical oxidation and AOPs [3, 15, 20] Adsorption [3, 4, 15, 20, 28–34] Membrane filtration [3, 4, 15, 20, 41] Ion exchange treatment [3, 4, 15, 25] Electrochemical treatment [3, 4, 15, 20]

3 Physical/Chemical Treatment of Leachate

In most leachate treatment plants, physical/chemical treatment is carried out as primary treatment prior to biological treatment. It is carried out to remove non-biodegradable materials including humic and fulvic acid, and other undesirable compounds such as heavy metals, adsorbable organic halogens and polychlorinated biphenyls from the leachate [15]. Sometimes, biological oxidation processes can be hindered by the presence of bio-refractory materials, which can be removed beforehand by physical/chemical system.

During physical treatment, only physical occurrence is utilized to improve leachate quality. There will be no chemical or biological changes in the leachate. Examples of physical treatment include screening to retain larger impurities, sedimentation to settle solids by gravitational force and aeration which utilizes oxygen as oxidation agent in removing BOD_5 [20]. Filtration, the simplest physical treatment, is used to trap and separate solid by passing leachate through a filter media.

On the other hand, chemical treatment consists of using chemical reactions to enhance leachate quality. The most commonly used chemical process is neutralization, where acid or base is added to adjust the pH of leachate back to neutrality. Coagulation consists of adding coagulant that will form an insoluble end product that serves to remove substances from leachate [20]. Adsorption involves both physical and chemical processes by using activated carbon to adsorb organics and metals from leachate.

Kumari et al. [26] studied the possibility of leachate treatment with microalgae and bacteria. Bacto-algal co-culture was observed to be most effective in removing toxic organic contaminants and heavy metals such as Zn, Cr, Fe, Ni, Pb, naphthalene, benzene, phenol and their derivatives, halogenated organic compounds naphthols, phthalates, pesticides and epoxides. Results of the study demonstrate the potential of bioremediation and detoxification of bacto-algal co-culture for leachate treatment through a significant decline in both cytotoxicity and genotoxicity.

Magnetic graphene oxide (MGO) coupled characters of graphene oxide and magnetism were used for removing humic acid/fulvic acid HA/FA and lead Pb(II) in landfill leachate. Maximum adsorption capacities reached for FA, HA and Pb(II) were 72.38, 98.82 and 58.43 mg/g, respectively [27].

Adsorption is commonly applied to remove organics and metals in leachate. Granular or powdered activated carbon has frequently been used due to its properties such as larger surface area, higher adsorption capacity and better thermal stability. Cui et al. [28] studied the effects of biological activated carbon (AC) dosage on COD removal in landfill leachate treatment. The COD removal efficiency was 12.9, 19.6 and 27.7% for reactors with 0, 100 and 300 g activated carbon dosage per litre activated sludge, respectively. Kaur et al. [29] used activated cow dung ash (ACA) and cow dung ash (CA) as adsorbents and obtained up to 79 and 66% removal of COD, respectively. Results indicated that ACA exhibited 11–13% better removal efficiency as compared to CA. Adsorption using granular activated carbon (GAC) by Kawahigashi et al. [30] obtained removals ranging between 94 and 100% for true colour, between 45 and 76% for COD and between 23 and 67% for TOC. Effluent from the Fenton process was subjected to adsorption with Carbotecnia lignite macroporous activated carbon and achieved 29 mg/L COD, 24 mg/L BOD_5 and 18 Pt–Co colour [31].

Gracilaria sp. extract was used for the treatment of matured landfill leachate containing As, Fe, Ni, Cd and formaldehyde. Results showed that high adsorption of heavy metals was observed with Fe absorbed at the fastest rate as compared to adsorption of formaldehyde [32]. Martins et al. [33] studied ammonium nitrogen adsorption followed by zeolite regeneration via nitrification of raw leachate. Zeolite initial adsorptive capacity was first evaluated, and then biological regeneration was performed using zeolite holding adsorbed ammonium in the presence of sodium bicarbonate and a nitrifying bacterial suspension. The adsorptive capacity was reduced by only 4.55% after regeneration from 11 to 10 mg NH_4^+–N/g zeolite.

In another adsorption study, cockleshells were used as an adsorbent media for the treatment of a stabilized landfill leachate. The optimum dosage, shaking speed and pH for COD removal were investigated using cockle shells of particle sizes ranging from 2.00 to 3.35 mm. Results showed that optimum shaking speed is 150 rpm and optimum pH and dosage was 5.5 and 35 g/L, respectively, according to the adsorption of COD. The adsorption isotherms suggested that a Langmuir isotherm was a better fit than a Freundlich isotherm [34].

Coagulation and flocculation is still the leading option for the treatment of landfill leachate and widely practiced by many landfill operators. In view of that, it is assumed that pre-hydrolysed coagulants can be considered as good coagulants due to better colour removal and high pH affinity towards leachate. Development of starch-based coagulants in leachate treatment is seen as a promising approach that can reduce the dependence of pure coagulant, which can effectively improve sludge production during the flocculation process. Still, the applicability of the starch-based coagulants is yet to be documented and extensively discussed [35]. Mechanically treated starch from oil palm trunk waste had been tested by Zamri et al. [36] as coagulant for semi-aerobic landfill leachate treatment. Oil palm trunk starch was found to be a better coagulant compared to polyaluminium chloride (PAC) in terms of COD removal, however, not effective for SS removal as compared to PAC.

In another study, a combination of a coagulation and nanofiltration process to treat leachate was performed. Poly ferric sulphate (PSF) was used as coagulant, and the results indicated that 62.8% COD and 75.3% turbidity removal efficiency can be obtained at an optimum dosage of 1200 mg/L at pH 6.0. In the nanofiltration process, 89.7% of COD, 78.2% of TOC, 72.5% of TN, 83.2% of TP and 78.6% of NH_3–N were retained when tested at 0.6 MPa at 25 °C. The final leachate effluent concentration reached for COD, BOD_5, NH_3–N, TOC and SS was 92, 31, 21, 73 and 23 mg/L, respectively [37].

A study by Dolar et al. [38] was carried out to evaluate a pretreatment alternative for treating stabilized landfill leachate using nanofiltration (NF) and RO through combination of coagulation/ultrafiltration (UF) and adsorption/UF. Coagulation showed better reduction of COD (65%), turbidity (87%) and TOC (86%) than adsorption. UF provided better results after adsorption since COD and TOC removal was higher, ratifying that larger molecules were removed with coagulation while smaller molecules removed with adsorption. Ammonium reduction was insignificant for all pretreatment steps.

Ishak et al. [39] studied the removal of H_2O_2 residue from stabilized landfill leachate that has undergone coagulation–flocculation coupled with the Fenton reaction. The highest TOC removal of 71% at pH 6 was achieved after coagulation–flocculation with ferric chloride. Almost 50% of TOC removal was achieved after the pretreated leachate was subjected to the Fenton reaction at pH 3, H_2O_2:Fe^{2+} ratio of 20:1, H_2O_2 dosage of 240 mM and 1 h of reaction time. The combination of coagulation–flocculation with the Fenton reaction removed 85% of TOC, 84% of COD and 100% turbidity. The ecotoxicity study performed using zebrafish revealed that the 96 h lethal concentration LC_{50} for raw stabilized leachate was 1.40% (v/v). After coagulation–flocculation, LC_{50} of the leachate was increased to 25.44%. However, after the Fenton reaction, the LC_{50} was found to decrease to 10.96% due to the presence of H_2O_2 residue. H_2O_2 residue was then removed by adsorption using powdered activated charcoal which successfully increased the LC_{50} of treated effluent to 34.48% and the removal of TOC and COD was further increased to 90%.

Biochar produced from palm oil mill effluent sludge was used to treat landfill leachate with the removal percentage of 44.26% of COD, 70.20% of colour and 30.56% of NH_3–N under optimal conditions. The optimization process of biochar preparation was done using RSM and the optimum conditions obtained were 452 °C pyrolysis temperature and 66 min of holding time [40].

Membrane processes have been utilized as an advanced type of leachate treatment. Combination of forward osmosis (FO) with membrane distillation (MD) is used to treat high salinity hazardous waste landfill leachate. Results presented showed that optimal feed solution flow rate, draw solution concentration and draw solution flow rate for FO stage were 0.87 L/min, 4.82 M and 0.31 L/min, respectively. Salt, TOC and TN rejection rates were higher than 96, 98 and 98%, respectively, and NH^{4+}–N, Hg, As and Sb were also successfully removed. This study proved that optimized FO-MD combined system can be used to treat high salinity hazardous wastewater [41].

Ozonation has also been applied in leachate treatment. A high removal efficiency of colour from concentrated leachate was reported by Mojiri et al. [42] when using a combination of electro-ozonation and a composite adsorbent augmented sequencing batch reactor (SBR) process. 64.8% COD, 90.4% colour and 52.9% nickel were removed at optimum reaction time of 96.9 min, pH 7.3, ozone dosage of 120.0 mg/L, current of 4 A and voltage of 9 V. Ozone consumption ranged from 0.3 to 1.4 kg COD removed per kg ozone. Concentrated leachate treated by electro-ozonation was further treated using the powdered composite adsorbent (P-BAZLSC) augmented SBR reactor (PB-SBR). PB–SBR enhanced the removal efficiency for COD, colour and nickel from 64.8 to 88.2%, 90.4 to 96.1% and 52.9 to 73.4%, respectively.

Single ultrasound (sonolytic) and sonolytic combined with Fe^{2+} and TiO_2 catalysts was utilized by Kocakaplan et al. [43] to treat leachate. The colour removal efficiency was recorded as 81.81% at 620 nm, pH = 2.0 and 70% wave amplitude. As the Fe^{2+} concentration increased from 1.0 to 3.0 mg/L, the COD and colour removal decreased from 50 to 35.7% and 95.5 to 90.9%, respectively. A combination of catalyst and sonolysis was shown to be effective to remove colour, COD and TOC from landfill leachate. Fenton-like with zero-valent iron (ZVI) process, an advanced oxidation process, was also investigated in treating leachate. The effect of pH on colour and COD removal from leachate was determined at initial pH values of 2.0, 2.5, 3.0, 3.5 and 4.0. The maximum colour and COD yield were achieved at pH 2.0 with colour and COD removal of 87.9% (in 620 nm wavelength) and 74%, respectively [44].

Treatment of leachate by a photocatalytic process using tungsten-doped TiO_2 nanoparticles under the fluorescent light irradiation was studied by Azadi et al. [45]. Calcination temperature, dopant content, pH and contact time of leachate with nanoparticles were first determined before concentration of the leachate COD was predicted using a quadratic regression equation obtained from RSM. Results showed at optimal conditions of pH 6.63, tungsten content of 2.64% by weight, contact time of 34 h and calcination temperature of 472 °C, 46% of COD was efficiently removed.

Electrolysis is also used to remove organic and inorganic pollutants in landfill leachate treatment. Electric potential, sodium chloride (NaCl) concentration, pH, hydraulic retention time (HRT), distance between electrodes and electrode materials were considered in a study. An electrical potential of 60 V at 5% of NaCl solution using Fe as cathode and Al as anode that was being kept at a distance of 3 cm with HRT of 120 min was the optimum operational condition for electrolysis resulting in high removal efficiencies of TSS, TDS, heavy metals, turbidity, salinity, BOD and COD. 94% COD removal and 93% Mn removal suggested that electrolysis is an efficient technique for multi-pollutant removal from landfill leachate [46].

Photoelectrocatalytic oxidation of landfill leachate using a Cu/N co-doped TiO_2 (Ti) electrode was investigated by Zhou et al. [47]. The experimental design method was applied to RSM for optimization of the operational parameters of the photoelectrocatalytic with TiO_2 as photoanode to treat landfill leachate. The results of the investigation revealed that 67% of COD and 82.5% of TOC were removed at optimum conditions for the degradation which are 4377.98 mg/L initial COD concentration, initial pH 10.0 and 25.0 V potential bias. Moreover, 38 out of 73 organic micro-pollutants disappeared completely in the photoelectrocatalytic process as shown by GC/MS.

Kim and Ahn [48] studied the effects of microwave-assisted persulfate oxidation (MAPO) on specific ultraviolet absorbance at 254 nm wavelength (SUVA254), absorbance spectra and colour in landfill leachate treatment. The study also looks at effects of treatment temperature and sodium persulfate (SPS) concentration. An absorbance band in the visible region disappeared, whereas the absorbance in UV region remained after MAPO. From the results, colour number (CN) (CN = 0.004/cm) removal of 99% was reached within 30 min with MAPO at 80 °C and 0.2 M SPS concentration. SUVA254 of treated leachate was found to vary from 2.3 to 14.7 L/mg m. In conclusion, as SPS concentration and temperature increased, SUVA254 also increased, however, TOC and CN decreased.

Microbial fuel cells (MFCs) are another type of advanced treatment for leachate. In a design by Alabiad et al. [49], activated carbon, carbon black and zinc electrodes used as anodes were tested in MFC reactors. Static and dynamic modes were studied to look at efficiency of MFC in eliminating ammonia. In both modes, activated carbon performed superior to zinc and black carbon. Removal rates of ammonia obtained were 96.6, 66.6 and 92.8% for activated carbon, zinc and black carbon.

A bioelectrochemical system (BES) is another promising technology for the concurrent removal and recovery of resources such as water, nutrients, energy and heavy metals. Organic compounds in the leachate undergo oxidation by microorganisms, thus producing energy and other valuable resources. High-quality water can be recovered through the integration of forward osmosis in BES. Uncertainty in concentration largely affects the recovery of metals from leachate. Ammonia and phosphorus recovery is through cathode reduction reactions driven by electricity generation. Among challenges for BES are low bioavailability of landfill leachate and system scaling up as energy production and consumption balance requires better understanding [50].

In general, physicochemical treatment processes are capital- and energy-intensive since cost of oxidants, membranes and chemical needs to be considered [51]. By looking at the results obtained from studies carried out using physical/chemical treatment of leachate as discussed above, it is apparent that physical/chemical treatment is effective in removal of COD, colour and heavy metals; however, it is not quite an effective option to treat ammonia nitrogen. Therefore, biological treatment is applied to further improve the quality of the leachate.

4 Biological Treatment of Leachate

Biological treatment utilizes microorganisms in the biochemical decomposition of leachate resulting in a more stable end product. Biologically treated leachate is almost totally free from biodegradable materials. Biological treatment methods can be divided into aerobic and anaerobic based on availability of dissolved oxygen. In aerobic condition, organic pollutants are mostly converted into CO_2 and sludge by using atmospheric O_2 transmitted to the wastewater. In anaerobic treatment, organic matter is transformed into biogas, mainly consisting of CO_2 and CH_4 and in a minor part into biological sludge. Biological processes are proven to remove nitrogenous and organic matter from young leachates effectively when the BOD/COD ratio has a high value, exceeding 0.5 [15]. Usually, a combination of aerobic/anaerobic system has been used to treat leachate more efficiently for the removal of biodegradable compounds.

Both aerobic and anaerobic processes have their own benefits and drawback. The aerobic-activated sludge process, for instance, requires high capital investment, and management and operation of the process are affected by temperature. Stabilization pond, on the other hand, requires a lengthy residence time (10–30 days), needs a large surface area and the purification capacity indicates large seasonal variations. Anaerobic treatment processes are suitable for high-strength organic wastewater. The disadvantages include a long retention time, and contaminant removal is quite low and more profound to temperature change. Anaerobic–aerobic biological treatment process of leachate is preferable, but management, operation and investment of the construction of the leachate treatment plant are costly, and once the landfill stop operation, the treatment facilities are eventually abandoned [9].

Anaerobic digestion involves the biological decomposition of inorganic and organic matter without the presence of molecular oxygen and as a result, CH_4 and CO_2 are produced. Anaerobic treatment has been normally used in tropical countries as a pretreatment since it requires less energy consumption, less sludge production and is excellent in treating high-strength wastewaters like leachate [52, 53]. Anaerobic process usually performs excellently under high organic loading rates, producing less residual sludge and in addition to that, generating biogas for renewable energy recovery [54].

Anaerobic treatment methods usually perform better than aerobic processes for biological treatment of landfill leachate due to its high COD content and high COD/BOD ratio. Using upflow anaerobic sludge blanket (UASB) reactors, up to 92% COD removal efficiency can be achieved. Anaerobic and sequential anaerobic–aerobic reactors used to treat leachate at temperatures between 11 and 24 °C resulted in removal efficiency of 80–90% for COD and nearly 80% for ammonium [18].

Anaerobic sequencing batch reactors (AnSBR) have been studied and considered a promising alternative to treat wastewater. The advantages of AnSBR are the probability to achieve high solids retention, offer better effluent quality control, high organic matter removal efficiency and the opportunity of appropriate process control [55]. At specific loading and volumetric rates varying from 0.17 to 1.85 g COD/g VSS/day and 0.4 to 9.4 g COD/L day, respectively, 64–85% COD removals can be achieved. Around 83% of COD removed is converted to CH_4 [56]. In a study by Kennedy and Lentz [57], at OLRs between 0.6 and 18.4 g COD/L day, the AnSBR removal efficiencies of soluble COD are between 71 and 92% while the continuous UASB reactor had soluble COD removal efficiency between 77 and 91%. Results from a study by Contrera et al. [52] showed more than 70% COD removal efficiency, with an inlet COD about 11,000 mg/L and a TVA/COD ratio of approximately 0.6 suggesting that the anaerobic sequence batch biofilm reactor (AnSBBR) could be a substitute for landfill leachate anaerobic pretreatment.

The problems of low growth rate of anaerobic microorganisms and the poor retention of biomass can be solved by integrating membrane separation with an anaerobic bioreactor to form the anaerobic membrane bioreactor (AnMBR). A dynamic membrane has been established, with advantages of comparable biomass retention efficiency but higher membrane flux and inexpensive membrane which can lessen the problem of membrane fouling. The anaerobic dynamic membrane bioreactor AnDMBR process achieved removal efficiency of 62% COD for raw leachate with COD of 13,000 mg/L. The CH_4 content in the biogas was within 70–90%, and at an OLR of 4.87 kg COD/(m^3 day), the average CH_4 yield was 0.34 L/g COD removed [58]. In another study, a lab-scale AnMBR was used to investigate the removal of five selected pharmaceutical compounds present in synthetic sewage. With addition of powdered activated carbon (PAC) to the AnMBR, the removal of all five compounds improved, with very high removal efficiencies for sulfamethoxazole (96%) and triclosan (93%) [59].

Anaerobic digestion can be employed at relatively low cost and it can generate an energy-rich biogas as a by-product [60]. Anaerobic digestion is a process suitable in handling large organic loads that usually characterize young leachates. One of the problems of using anaerobic digestion in leachate treatment is slow growth of microorganisms; hence, a conventional anaerobic digester requires a huge digester volume, which was expensive. In addition, high flow rate in the conventional anaerobic digester always lead to wash out. Therefore, media is used to halt the bacteria inside a fixed bed anaerobic digester to stabilize and maximize the microorganism's growth and also reducing the likelihood of the bacteria to be washed out. Natural zeolite is one possible material for the immobilization media that was able to adsorb ammonia produced during the leachate digestion [61].

The membrane bioreactor (MBR) is another alternative to treat old landfill leachate. The membrane separation capacity of an MBR could retain most microbial cells in the reactor to sustain a high biomass concentration, making it an effective biological digestion system. As compared to conventional biological systems, more than 90% BOD and ammonia removals as well as 75% or higher COD removal can be achieved with larger organic loading rates (OLR) and shorter hydraulic residence times (HRT). Latest developments such as AnMBR and PAC-amended MBR have shown great capabilities in treating landfill leachate [62]. Application of SMBR resulted in high BOD and P removal and nitrification. Due to strong interaction with negative surface charge of sludge, MBR performance was excellent for heavy metals such as Fe, Cu, Al, Zn, Pb, Cd and Cr but not for Mo, As, Mn and Ni [63].

In investigating anaerobic digested effluent treatment performance and mechanisms, Kizito et al. [64] comparatively evaluated three types of vertical flow constructed wetland columns (VFCWs), packed with corn cob biochar (CB-CW), wood biochar (WB-CW) and gravel (G-CW) under tidal flow operations. As result, CB-CW and WB-CW provide significantly higher removal efficiencies for organic matter (>59%), NH_4^+–N (>76%), TN (>37%) and phosphorus (>71%), compared with G-CW (22–49%). The higher pollutant removal ability of biochar-packed VFCWs was mainly attributed to the higher adsorption ability and microbial cultivation in the porous biochar media. Increasing the flooded/drained ratio from 4/8 to 8/4 h of the tidal operation further improved around 10% of the removal of both organics and NH_4^+–N for biochar-packed VFCWs. The study showed that the use of biochar would enhance the treatment performance and lengthen the lifespan of CWs under tidal operation.

In another study, Kizito et al. [65] used slow pyrolyzed biochars produced from wood (WDB), corncobs (CCB), rice husks (RHB) and sawdust (SDB) for adsorption, desorption and regeneration of phosphate (PO_4^{3-}-P) from anaerobically digested liquid swine manure. The PO_4^{3-}–P adsorption capacity increased followed by initial concentrations increasing. Pseudo-second-order kinetics model could best fit PO_4^{3-}–P adsorption; therefore, indicating the chemisorption via precipitation was the main mechanism for PO_4^{3-}–P removal. This finding suggested biochar could be effectively used to recover PO_4^{3-}–P from anaerobic digestate. Previously, Kizito et al. [66] have done a study that demonstrated the potential of wood biochar and rice husk to adsorb NH_4^+–N from piggery manure anaerobic digestate slurry. The treated slurry can be used as nutrient filters before released into water streams. The sorption process was endothermic and followed pseudo-second-order kinetic models (R^2 = 0.998 and 0.999) and Langmuir (R^2 = 0.995 and 0.998) models.

Almost 99% NH_4–N removal was achieved after undergoing aerobic treatment of domestic leachates in a sequencing batch reactor (SBR) with 20–40 days of residence time [67]. Nearly, 95% BOD and 50% nitrogen were removed from combined landfill leachate and domestic sewage that was treated in an SBR with the ratio of sewage to leachate 9:1 [68]. Mojiri et al. [69] added powdered ZELIAC (PZ), which is an adsorbent consisting of zeolite, activated carbon, limestone, rise husk ash and Portland cement to the SBR to study the treatment of landfill leachate and domestic wastewater. PZ–SBR showed that better performances with removal efficiencies for colour, phenols, COD and NH_3–N were 84.11, 61.32, 72.84 and 99.01%, respectively, at optimum conditions of leachate to wastewater ratio (20%), aeration rate (1.74 L/min) and contact time (10.31 h) [69]. In another study, PZ was used to remove heavy metals from landfill leachate and urban wastewater. PZ–SBR removed heavy metals more efficiently than SBR. In PZ–SBR, removal efficiencies for Fe, Mn, Ni and Cd were 79.57, 73.38, 79.29 and 76.96%, respectively, at optimum contact time (11.70 h), leachate to wastewater ratio (20.13%) and aeration rate (2.87 L/min) [70].

For large-scale application, it is substantial to combine aged refuse reactor and slag reactor [71]. Using aged refuse under the conditions of 700 °C, pH 9, aged refuse dosage of 60 g/L and reaction time of 10 h, the COD and ammonia nitrogen removal rates reached 58.38 and 79.77% and the equilibrium adsorption capacity were 11.68 and 1.58 mg/g. The fitting of dynamic data from the COD and ammonia nitrogen adsorption processes was in line with the quasi-second-order equation, which indicated that the rate of adsorption was dependent on chemical adsorption. The fitting of the isotherm equation showed that the adsorption of COD and ammonia nitrogen to aged refuse can be classified as multilayer adsorption and monolayer adsorption, respectively [72].

There are many reports on practical methods for leachate treatment operating under anaerobic or aerobic conditions [52, 73, 74]. It has also been considered using natural systems like artificial wetlands and ponds in design of treatment facilities [75], whereas evaporation and leachate recirculation are discussed in other systems [76, 77].

In a study, methanogenesis and denitrification were performed in the anaerobic reactor, whereas nitrification of NH_4^+–N and organic removal were carried out in the aerobic reactor. In the anaerobic reactor, the maximum organic removal rate was 15.2 kg COD/m^3 day while in the aerobic reactor, the maximum nitrification rate and maximum NH^{4+}–N removal rate were 0.50 kg NO^{3-}–N/m^3 day and 0.84 kg NH^{4+}–N/m^3 day, respectively. For proper nitrification in the aerobic reactor, the pH range was 6–8.8 [78].

Anaerobic treatment of municipal landfill leachate was studied using laboratory-scale UASB and hybrid reactors at 11 and 24 °C. The anaerobically treated leachate was subjected to aerobic post-treatment using an activated sludge process, also at 24 °C. At 1.5–2 day HRT and 0.7–1.5 kg COD/m^3 day organic loading rates in the 11 °C reactors, up to 60–65% COD removal was obtained, whereas with a 10 h HRT at 24 °C, up to 75% COD removal was achieved. 45–75% of the COD left after anaerobic treatment was removed by the aerobic post-treatment, thus producing effluent with a BOD_7 of less than 22 mg/L and less than 380 mg/L COD. In the sequential process, the COD removal was 80–90%. In the aerobic stage, up to 80% ammonium and 40% total nitrogen were removed [79]. In another study using a UASB reactor, the COD removal efficiency achieved a maximum of 80% as the OLR increased from 4.3 to 16 kg/m^3 per day. After the aerobic stage, NH_4–N removal efficiency was around 99.6% [80].

In a system consisting of a UASB, anoxic/aerobic (A/O) reactor, and a denitrifying UASB (DUASB), a mixture of 1:2 raw leachate and domestic wastewater treated. The total nitrogen, organic matter and ammonia nitrogen in the final effluent were 39, 80–90 and 14 mg/L, respectively, with respective removal efficiencies of 98, 95 and 95%. COD removal efficiencies in the UASB, A/O and denitrifying UASB were 76.8, 2.8 and 16.3%, respectively [81].

In a batch reactor, anoxic digestion based on the endogenous biomass activity has been studied. The anoxic digestion reactor has shown 91% BOD_5 reductions, as well as 46, 65, 45 and 63% reduction for COD, TOC, NH_4–N and TKN, respectively, with a retention time of 90 days. With 7 days of total retention time, the effluent was then treated in downflow cascade in three aerated submerged biological reactors. In the aerobic reactors, further reductions were achieved and overall removal efficiencies of 95, 94 and 92% for BOD_5, COD and NH^{4+}–N, respectively, were achieved by the coupled system of anoxic and aerobic reactors [82].

5 Combination of Physical/Chemical Treatment with Biological Treatment

In larger systems and depending on the treatment goals, integrated systems which combine the physical/chemical and biological processes are often used. Removal efficiency of contaminants can be increased as a result of these integrated systems.

In a study by Bakraouy et al. [83], landfill leachate is treated anaerobically combined with a coagulation–flocculation process, using ferric chloride as coagulant and a cationic polymer as flocculant. Optimal dosages were 4.4 g/L of coagulant and 9.9 mL/L of flocculant. Removal efficiencies reached: 89, 69, 94, 80 and 89% for phenol, turbidity, colour, COD and absorbance at 254 nm, respectively.

Ammonia removal ranging from 82 to 93% and average COD and BOD removal of 64 and 67%, respectively, were achieved when combining air stripping with aerobic biological treatment in treating leachate. Meanwhile, respective removal values of COD and BOD for anaerobic treatment were 41 and 45.5% with no reduction of ammonia concentration recorded. Using the same operating conditions, an evaluation between aerobic and anaerobic treatment has a preference for ammonia stripping followed by aerobic treatment [84].

In a study of sanitary landfill leachate treatment, a combination of biofiltration (BF) and electrocoagulation (EC) processes was investigated. $N–NH_4$, BOD, turbidity and phosphorus removal of 94, 94, 95 and >98%, respectively, was obtained when BF process was used as secondary treatment. For tertiary treatment, EC process using magnesium-based anode was used. The best performances were achieved by applying a current density of 10 mA/cm^2 through 30 min of treatment with 53% COD removal and 85% colour removal recorded [85].

Highly contaminated old landfill leachate was treated using combination of high-performance MBR equipped with UF and electro-oxidation process (EOP) by boron-doped diamond electrode (BDD). MBR and EOP were optimized for both raw and pretreated landfill leachates. Sludge retention time of 80 days and organic load rate of 1.2 g COD/L day was considered as the optimum operating condition for MBR in which respective removal efficiencies of COD, NH^{4+}, TOC and P attained the average of 63, 98, 35 and 52% [86].

In another treatment of landfill leachate, a combined process of anaerobic digestion, lime precipitation, microfiltration (MF) and reverse osmosis was studied. The OLR was increased gradually up to 3.3 g COD/L day during the anaerobic digestion step. The upflow anaerobic fixed bed reactor presented excellent efficiency in terms of biogas production and COD removal. Lime dose was enhanced to attain maximum reduction of conductivity during precipitation experiments to avoid RO membranes fouling. Similar removal efficiencies were obtained when anaerobic digestion step was removed. Considerably, the anaerobic digestion step enhanced the process by reducing 50% of the lime dose and increasing 35% of MF and 40% of RO fluxes at a steady state [87].

Based on results of studies discussed above, biological treatment proves useful in treating raw or pretreated leachate with high concentrations of organic substances. However, the success of the treatment drops as the landfill age increases, due to reduction of biodegradable organic matter over time and leachate turns into stabilized leachate [88]. Most of the old or biologically treated leachate contains high amount of recalcitrant organic molecules that cannot be removed totally using biological treatment [15]. Further, post-treatment is deemed necessary in order to meet stringent quality standards for discharge of leachate into the receiving water.

6 Adsorption as Post-Treatment of Biologically Digested Leachate

Biological treatment has a poor removal efficiency for certain substances such as halogenated bio-refractory organic compounds (AOX) and metals [88]. Because of this, the organic components present in stabilized landfill leachate were mostly non-biodegradable. Moreover, the presence of high-strength NH_3–N and bio-refractory materials in the stabilized leachate usually may inhibit the biological activity in the bioreactor [3, 89]. As such, conventional biological treatment methods are no longer adequate and efficient to be utilized for stabilized landfill leachate treatment; therefore, physical/chemical treatments are recommended as a refining step for biologically digested leachate and also to remove refractory substances from stabilized leachate [20, 88].

Physical/chemical methods such as ozonation, coagulation–flocculation, Fenton oxidation, membrane filtration and activated carbon adsorption are regularly used for post-treatment of landfill leachate [4, 90–92]. Adsorption is reported as frequently used as post-treatment of biologically digested landfill leachate [10]. Although other adsorbents such as zeolites are used, the most common adsorbent being used for treating stabilized leachate is activated carbon. Activated carbon has large specific surface area, fast adsorption kinetics and thermostability. Activated carbon is also very flexible in removing wide range of organic and inorganic pollutants of various concentrations [93]. Some of the studies on adsorption of biologically digested leachate are as summarized in Table 5.

An aeration tank operated in fed-batch mode with and without powdered activated carbon (PAC) as adsorbent was used as post-treatment to treat pretreated leachate using coagulation–flocculation and air stripping. At concentrations above 0.5 g/L, PAC addition improved COD removal significantly but marginally for PAC concentrations above 2 g/L. With 2 g/L PAC added biological treatment, almost 86% COD removal was achieved. As compared to treatment using only biological oxidation and only PAC adsorption, COD removals were almost 74 and 38%, respectively, after 30 h fed-batch operation [18].

Adsorption process was performed by Rodríguez et al. [103] on final effluent from the La Zoreda landfill leachate treatment plant. Prior to adsorption, the leachate was recirculated through an anaerobic digestion pilot plant. Effluent resulting from the anaerobic had COD exceeding 1000 mg/L, indicating a residual non-biodegradable organic matter still remains. Amberlite XAD 4, Amberlite XAD 8, Amberlite IR-120 and GAC were adsorbents used for adsorption, and it was found that GAC could remove the highest COD, followed by Amberlite XAD-8. When using GAC, 200 mg/L of residual COD was obtained, while 59% of the initial COD, which may be related to fulvic acids, was removed by the resin Amberlite XAD-8.

Papastavrou et al. [98] studied the post-treatment of a biologically treated landfill leachate by comparing two treatment schemes which were coagulation followed by AC adsorption, and electrochemical treatment. 50% of COD was removed by

Table 5 Treatment effectiveness of biologically digested leachate using adsorption

Source of leachate	Pretreatment/Biological treatment	Adsorbent	Operating parameter		Performance/Efficiency (removal %)				References
			Dosage	System	COD	NH_4	Colour	Others	
Spain landfill	Pressurized nitrification–denitrification and UF	Organosorb 10 GAC, BET surface area: 1020 m^2/g	20 g/L	Packed bed	45		25		[94]
		Organosorb 10 MB GAC, BET surface area: 1020 m^2/g			63		45		
		Filtracarb CC65/1240 GAC, BET surface area: 1050 m^2/g			58		35		
India landfill	Sand filter, SBR with nitrification denitrification and coagulation	Coconut shell GAC	10 g/L		89.41	81.24	95.42	Total solids = 84.64	[95]
Stabilized Belgium landfill	Nitrification/Denitrification, ion exchange	Desotec GAC, porosity 0.83		Glass column	268 mg/g max adsorption capacity			AOXs = 0.58 mg/g max adsorption capacity	[96]
Shanghai landfill	3 stage Age Refused Bioreactor (ARB)	Commercial PAC, 300 mol sieve	3 g/L		24.6			HOCs = 89.2	[97]
		Commercial GAC, 40 mol sieve	3 g/L		19.1			HOCs = 73.4	
		Biomimetic fat cell (BFC), specific area: 30.2755 m^2/g	3 g/L		8.9			HOCs = 81.1	
Greece landfill	Sequencing batch reactors, constructed wetlands and coagulation	GAC (YAO M200 W20) and PAC (YAO 30 × 60)			80				[98]

(continued)

Table 5 (continued)

Source of leachate	Pretreatment/Biological treatment	Adsorbent	Operating parameter		Performance/Efficiency (removal %)				References
			Dosage	System	COD	NH_4	Colour	Others	
Tunisia landfill	Anoxic digestion and aerated submerged biological reactors	PAC			99.7				[82]
Stabilized Belgium landfill	Biological treatment	Commercial GAC		Column bed	20			$\alpha_{254} = 8$	[21]
	Biological treatment and coagulation–flocculation ($FeCl_3$)				53			$\alpha_{254} = 99$	
	Biological treatment and ozonation				77			$\alpha_{254} = 90$	
Stabilized Belgium landfill	Sequencing batch reactor (SBR) and UV/H_2O_2 treatment	GAC (Organosorb10)			6.9				[99]
	Sequencing batch reactor (SBR) and ozonation				10				
	Sequencing batch reactor (SBR) and photo-Fenton				63				
Malaysia landfill	Aerobic sequencing batch reactor	Zeolite, particle size 2.5–5.0 mm	10		43	96		Aluminium = 100 Vanadium = 44 Chromium = 63 Magnesium = 75 Cuprum = 24 Plumbum = 85	[100]
UK landfill	Aerobic process and UF	Pulverized coal-based GAC	10 g/L		80			TOC = 80 Heavy metals = 80	[101]
Municipal landfill	MBR and AOP	Sludge-derived biochar (SDBC), GAC surface area: 75.34 m^2/g	5 g/L			98.7		TOC = 56	[102]

coagulation with alum at an optimum dose of 3 mM Al^{3+}, while an overall of 80% of COD was removed after AC adsorption. 170 mg/L of the organic matter was not adsorbable. Nearly, 90% COD removal in 240 min was obtained after electrochemical oxidation over a boron-doped diamond electrode.

Study by Oloibiri et al. [21] presented the advantage of using combined treatment trains to treat biologically stabilized leachate, adding an ozonation or coagulation step before GAC could enhance organic matter removal from the leachate. By applying $FeCl_3$ coagulation–flocculation before GAC adsorption, an overall removal of 53% COD and 99% α_{254} was reached. Ozonation/GAC combination sequence resulted in up to 77% COD removal.

Another study looked at adsorption using AC to treat biologically treated leachates followed by UF. Two types of pulverized coal-based granular activated carbon were used for the study. 80% removal of COD and TOC are attained by increasing carbon dose, and at a dose of 10 g/L of pulverized AC, almost similar removal rates for each carbon product were achieved. More than 50% Cu and about 65% Ni was removed by doses of 10 g/L of pulverized AC [101]. Post-treatment by adsorption on PAC to remove heavy metals was also studied by Trabelsi et al. [82] which attained total reduction level of ~99% of COD.

Gao et al. [104] looked at economical yet effective nitrogen and COD removal from biologically treated leachate using the following post-treatment method: ozonation, adsorption to GAC, ozonation followed by GAC and recirculating ozonated effluent into autotrophic nitrogen removal (ANR) system. Prior to post-treatment, the leachate was biologically treated with the ANR process. Only 14% of total nitrogen and 15% residual COD in the leachate can be removed by ANR post-treatment with ozonation, whereas 17% of total nitrogen and 74% of COD was effectively removed using AC. An overall removal performance of 83% for COD and 78% for nitrogen was attained using combination of ANR, ozonation and GAC.

Multiple oxidation processes, namely UV/H_2O_2 treatment, ozonation and photo-Fenton treatment, were compared to treat biologically stabilized leachate prior to GAC filtration. Although oxidation can modify the affinity of the organic matter towards GAC, adsorption using GAC (Organosorb10) presented uncertain change of adsorption capacity. Initially, a decreased adsorption onto GAC happened as result of polarity changes from a hydrophobic to more hydrophilic character. Then, higher adsorption capacity was achieved as larger components were partially converted into smaller components that could infiltrate deeper and easier into the pores of the GAC. No change in adsorption was observed as both effects appear to neutralize each other. 6.9, 10 and 63% COD removal were achieved after GAC adsorption pretreated by UV/H_2O_2 treatment, ozonation and photo-Fenton treatment, respectively [99].

Combined adsorption and coagulation was studied to treat a biologically treated leachate from an industrial landfill. By addition of 490 mg alum/L and 1000 mg PAC/L in an adsorption–coagulation process with pH control, up to 32 and 68% of the COD and colour were removed, respectively. Adsorption and coagulation combination process with pH control can remove COD and colour better as compared to process without pH control. COD removal was not influenced by pH control, while colour removal was influenced greatly by pH control [105].

Adsorption was investigated as a post-treatment to biologically treated leachates produced at the MSW landfill of Asturias (Spain). Adsorbents used were different AC: Organosorb 10MB, Organosorb 10 and Filtracarb CC65/1240. With Organosorb 10MB, adsorption capacities ranged between 150 and 157 mg COD/g were obtained for an AC dosage of 1 mg/L, while adsorption capacities between 13.3 and 18.4 mg COD/g were achieved for an AC dosage of 20 mg/L. Adsorption capacities ranged between 145 and 175 PtCo/g for the lower dosage and between 16 and 29 PtCo/g for the higher dosage are obtained for colour. Highest COD and colour removals of 63 and 45%, respectively, were achieved for a dosage of 20 mg/L after 5 h contact time, suggesting removal efficiency increased as the dosage of AC increased [94].

Based on results obtained from the studies discussed above, adsorption process has shown its effectiveness in removing ammonia nitrogen and metals from biologically digested leachate. Removal effectiveness of COD by adsorption depended on type of other pretreatment that was carried out prior to adsorption process. Removal of colour is speculated to be mainly assisted by coagulation process. This can be supported through findings by Hur et al. [106] in a study that COD removal was achieved mainly by adsorption, and colour was removed mainly by coagulation in the adsorption–flocculation process. Non-biodegradable organics, COD and colour may also be reduced by adsorption process due to the presence of AC or other type of adsorbents, which is assumed as providing synergy effect for preparing a surface for attachment of bio-regeneration (microorganisms) and becomes a core for floc formation occurrence, as suggested by Çeçen et al. [107].

7 Conclusion and Future Work

Despite tremendous efforts performed by landfill operators and designers to elude the increasing volume of waste generation, through efforts of composting, reuse, reduce and recycling including other various approaches of waste pretreatment prior to landfilling, landfilling is expected to remain as the final destination of waste. In view of that, leachate will continue to be generated and its discharge after undergoing treatment will remain as a problem requiring serious attention. There are still numerous landfill sites that are either in operation or already closed, which will be generating leachate for ages. For that reason, the best technology or integration of technologies in treating leachate must be identified that preferably suits the conditions of effective, low cost, ease of operation and durability.

This chapter indicates that adsorption has proven to be promising in post-treatment of biologically treated leachate. However, most studies are using activated carbon as the adsorbent, be it in granular or powder form, which is costly in terms of operation and regeneration, resulting in increased treatment costs.

Opportunity lies in searching for a better type of adsorbent which is environmental friendly, effective, low cost and accessible such as biosorbent.

References

1. Aziz HA, Adlan MN, Zahari MSM, Alias S (2004) Removal of ammoniacal nitrogen ($N\text{-}NH_3$) from municipal solid waste leachate by using activated carbon and limestone. Waste Manage Res 22(5):371–375
2. Li W, Hua T, Zhou Q, Zhang S, Li F (2010) Treatment of stabilized landfill leachate by the combined process of coagulation/flocculation and powder activated carbon adsorption. Desalination 264(1):56–62. https://doi.org/10.1016/j.desal.2010.07.004
3. Gao J, Oloibiri V, Chys M, Audenaert W, Decostere B, He Y, Van Langenhove H, Demeestere K, Van Hulle SWH (2015) The present status of landfill leachate treatment and its development trend from a technological point of view. Rev Environ Sci Bio/Technol 14 (1):93–122. https://doi.org/10.1007/s11157-014-9349-z
4. Kurniawan TA, W-h Lo, Chan GYS (2006) Physico-chemical treatments for removal of recalcitrant contaminants from landfill leachate. J Hazard Mater 129(1):80–100. https://doi.org/10.1016/j.jhazmat.2005.08.010
5. Bashir MJK, Xian TM, Shehzad A, Sethupahi S, Choon Aun N, Abu Amr S (2017) Sequential treatment for landfill leachate by applying coagulation-adsorption process. Geosyst Eng 20(1):9–20. https://doi.org/10.1080/12269328.2016.1217798
6. Rui LM, Daud Z, Latif AAA (2012) Treatment of Leachate by coagulation-flocculation using different coagulants and polymer: a review. Int J Adv Sci Eng Inf Technol 2(2):114–117
7. Moody CM, Townsend TG (2017) A comparison of landfill leachates based on waste composition. Waste Manage 63:267–274. https://doi.org/10.1016/j.wasman.2016.09.020
8. Rafizul IM, Alamgir M (2012) Characterization and tropical seasonal variation of leachate: results from landfill lysimeter studied. Waste Manage 32(11):2080–2095. https://doi.org/10.1016/j.wasman.2012.01.020
9. Peng Y (2017) Perspectives on technology for landfill leachate treatment. Arab J Chem 10: S2567–S2574. https://doi.org/10.1016/j.arabjc.2013.09.031
10. Foo KY, Hameed BH (2009) An overview of landfill leachate treatment via activated carbon adsorption process. J Hazard Mater 171(1):54–60. https://doi.org/10.1016/j.jhazmat.2009.06.038
11. Alvarez-Vazquez H, Jefferson B, Judd SJ (2004) Membrane bioreactors vs conventional biological treatment of landfill leachate: a brief review. J Chem Technol Biotechnol 79 (10):1043–1049
12. Chian ESK, DeWalle FB (1976) Sanitary landfill leachates and their treatment. ASCE J Environ Eng Div 2(2):411–431
13. Gálvez A, Greenman J, Ieropoulos I (2009) Landfill leachate treatment with microbial fuel cells; scale-up through plurality. Bioresour Technol 100(21):5085–5091. https://doi.org/10.1016/j.biortech.2009.05.061
14. Tugtas AE, Cavdar P, Calli B (2013) Bio-electrochemical post-treatment of anaerobically treated landfill leachate. Bioresour Technol 128:266–272. https://doi.org/10.1016/j.biortech.2012.10.035
15. Abbas AA, Jingsong G, Ping LZ, Ya PY, Al-Rekabi WS (2009) Review on landfill leachate treatments. J Appl Sci Res 5(5):534–545
16. Bove D, Merello S, Frumento D, Arni SA, Aliakbarian B, Converti A (2015) A critical review of biological processes and technologies for landfill leachate treatment. Chem Eng Technol 38(12):2115–2126

17. Renou S, Givaudan JG, Poulain S, Dirassouyan F, Moulin P (2008) Landfill leachate treatment: review and opportunity. J Hazard Mater 150(3):468–493. https://doi.org/10.1016/j.jhazmat.2007.09.077
18. Kargi F, Yunus Pamukoglu M (2003) Simultaneous adsorption and biological treatment of pre-treated landfill leachate by fed-batch operation. Process Biochem 38(10):1413–1420. https://doi.org/10.1016/S0032-9592(03)00030-X
19. Keenan JD, Steiner RL, Fungaroli AA (1984) Landfill leachate treatment. J Water Pollut Control Fed 56(1):27–33
20. Kamaruddin MA, Yusoff MS, Aziz HA, Hung Y-T (2015) Sustainable treatment of landfill leachate. Appl Water Sci 5(2):113–126. https://doi.org/10.1007/s13201-014-0177-7
21. Oloibiri V, Ufomba I, Chys M, Audenaert WTM, Demeestere K, Van Hulle SWH (2015) A comparative study on the efficiency of ozonation and coagulation–flocculation as pretreatment to activated carbon adsorption of biologically stabilized landfill leachate. Waste Manage 43:335–342. https://doi.org/10.1016/j.wasman.2015.06.014
22. Azmi NB, Bashir MJK, Sethupathi S, Wei LJ, Aun NC (2015) Stabilized landfill leachate treatment by sugarcane bagasse derived activated carbon for removal of color, COD and NH3-N—optimization of preparation conditions by RSM. J Environ Chem Eng 3(2):1287–1294. https://doi.org/10.1016/j.jece.2014.12.002
23. Trebouet D, Schlumpf JP, Jaouen P, Quemeneur F (2001) Stabilized landfill leachate treatment by combined physicochemical–nanofiltration processes. Water Res 35(12):2935–2942. https://doi.org/10.1016/S0043-1354(01)00005-7
24. Singh S, Janardhana Raju N, RamaKrishna C (2017) Assessment of the effect of landfill leachate irrigation of different doses on wheat plant growth and harvest index: a laboratory simulation study. Environ Nanotechnol Monit Manage 8:150–156. https://doi.org/10.1016/j.enmm.2017.07.005
25. Bashir MJ, Aziz HA, Amr SSA, Sap Sethupathi, Ng CA, Lim JW (2015) The competency of various applied strategies in treating tropical municipal landfill leachate. Desalin Water Treat 54(9):2382–2395
26. Kumari M, Ghosh P, Thakur IS (2016) Landfill leachate treatment using bacto-algal co-culture: an integrated approach using chemical analyses and toxicological assessment. Ecotoxicol Environ Saf 128:44–51. https://doi.org/10.1016/j.ecoenv.2016.02.009
27. Zhang J, Gong J-L, Zenga G-M, Ou X-M, Jiang Y, Chang Y-N, Guo M, Zhang C, Liu H-Y (2016) Simultaneous removal of humic acid/fulvic acid and lead from landfill leachate using magnetic graphene oxide. Appl Surf Sci 370:335–350. https://doi.org/10.1016/j.apsusc.2016.02.181
28. Cui YR, Guo Y, Wu Q, Ma LD, Sun JH, Cui FL (2014) Influence of biological activated carbon dosage on landfill leachate treatment. Huanjing Kexue/Environ Sci 35(8):3206–3211. https://doi.org/10.13227/j.hjkx.2014.08.052
29. Kaur K, Mor S, Ravindra K (2016) Removal of chemical oxygen demand from landfill leachate using cow-dung ash as a low-cost adsorbent. J Colloid Interface Sci 469:338–343. https://doi.org/10.1016/j.jcis.2016.02.025
30. Kawahigashi F, Mendes MB, da Assunção Júnior VG, Gomes VH, Fernandes F, Hirooka EY, Kuroda EK (2014) Post-treatment of landfill leachate using activated carbon. Engenharia Sanit Ambiental 19(3):235–244. https://doi.org/10.1590/S1413-41522014019000000652
31. Ramírez-Sosa DR, Castillo-Borges ER, Méndez-Novelo RI, Sauri-Riancho MR, Barceló-Quintal M, Marrufo-Gómez JM (2013) Determination of organic compounds in landfill leachates treated by Fenton-Adsorption. Waste Manage 33(2):390–395. https://doi.org/10.1016/j.wasman.2012.07.019
32. Selvam SB, Chelliapan S, Din MFM, Shahperi R, Aris MAM (2016) Adsorption of heavy metals from matured leachate by Gracilaria. Sp extract. Res J Pharm Biol Chem Sci 7(3):15–20
33. Martins TH, Souza TSO, Foresti E (2017) Ammonium removal from landfill leachate by Clinoptilolite adsorption followed by bioregeneration. J Environ Chem Eng 5(1):63–68. https://doi.org/10.1016/j.jece.2016.11.024

34. Daud Z, Abubakar MH, Kadir AA, Latiff AAA, Awang H, Halim AA, Marto A (2017) Adsorption studies of leachate on cockle shells. Int J GEOMATE 12(29):2186–2990
35. Kamaruddin MA, Abdullah MMA, Yusoff MS, Alrozi R, Neculai O (2017) Coagulation-flocculation process in landfill leachate treatment: focus on coagulants and coagulants aid. IOP Conf Ser Mater Sci Eng 209(1):012083
36. Zamri MFMA, Yusoff MS, Aziz HA, Rui LM (2016) The effectiveness of oil palm trunk waste derived coagulant for landfill leachate treatment. In: AIP conference proceedings, 2016. https://doi.org/10.1063/1.4965073
37. Shu Z, Lü Y, Huang J, Zhang W (2016) Treatment of compost leachate by the combination of coagulation and membrane process. Chin J Chem Eng 24(10):1369–1374. https://doi.org/10.1016/j.cjche.2016.05.022
38. Dolar D, Košutić K, Strmecky T (2016) Hybrid processes for treatment of landfill leachate: coagulation/UF/NF-RO and adsorption/UF/NF-RO. Sep Purif Technol 168:39–46. https://doi.org/10.1016/j.seppur.2016.05.016
39. Ishak AR, Hamid FS, Mohamad S, Tay KS (2017) Removal of organic matter from stabilized landfill leachate using Coagulation-Flocculation-Fenton coupled with activated charcoal adsorption. Waste Manage Res 35(7):739–746. https://doi.org/10.1177/0734242X17707572
40. Bashir MJK, Wong JW, Sethupathi S, Aun NC, Wei LJ (2017) Preparation of palm oil mill effluent sludge biochar for the treatment of landfill leachate. In: MATEC Web of conferences, 2017. https://doi.org/10.1051/matecconf/201710306008
41. Zhou Y, Huang M, Deng Q, Cai T (2017) Combination and performance of forward osmosis and membrane distillation (FO-MD) for treatment of high salinity landfill leachate. Desalination 420:99–105. https://doi.org/10.1016/j.desal.2017.06.027
42. Mojiri A, Ziyang L, Hui W, Ahmad Z, Tajuddin RM, Abu Amr SS, Kindaichi T, Aziz HA, Farraji H (2017) Concentrated landfill leachate treatment with a combined system including electro-ozonation and composite adsorbent augmented sequencing batch reactor process. Process Saf Environ Prot. https://doi.org/10.1016/j.psep.2017.07.013
43. Kocakaplan N, Ertugay N, Malkoç E (2017) The degradation of landfill leachate in the presence of different catalysts by sonolytic and sonocatalytic processes. Part Sci Technol 1–8. https://doi.org/10.1080/02726351.2017.1297338
44. Ertugay N, Kocakaplan N, Malkoç E (2017) Investigation of pH effect by Fenton-like oxidation with ZVI in treatment of the landfill leachate. Int J Min Reclam Environ 31(6):404–411. https://doi.org/10.1080/17480930.2017.1336608
45. Azadi S, Karimi-Jashni A, Javadpour S (2017) Photocatalytic treatment of landfill leachate using W-Doped TiO_2 Nanoparticles. J Environ Eng (United States) 143(9). https://doi.org/10.1061/(asce)ee.1943-7870.0001244
46. Erabee IK, Ahsan A, Jose B, Arunkumar T, Sathyamurthy R, Idrus S, Daud NNN (2017) Effects of electric potential, NaCl, pH and distance between electrodes on efficiency of electrolysis in landfill leachate treatment. J Environ Sci Health Part A 52(8):735–741. https://doi.org/10.1080/10934529.2017.1303309
47. Zhou X, Zhou S, Feng X (2017) Optimization of the photoelectrocatalytic oxidation of landfill leachate using copper and nitrate co-doped TiO_2 (Ti) by response surface methodology. PLoS ONE 12(7):e0171234. https://doi.org/10.1371/journal.pone.0171234
48. Kim Y-B, Ahn J-H (2017) Changes of absorption spectra, SUVA254, and color in treating landfill leachate using microwave-assisted persulfate oxidation. Korean J Chem Eng 34(7):1980–1984. https://doi.org/10.1007/s11814-017-0104-3
49. Alabiad I, Ali UFM, Zakarya IA, Ibrahim N, Radzi RW, Zulkurnai NZ, Azmi NH (2017) Ammonia removal via microbial fuel cell (MFC) dynamic reactor. IOP Conf Ser Mater Sci Eng 206(1):012079
50. Shehzad A, Bashir MJK, Sethupathi S, Lim JW, Younas M (2016) Bioelectrochemical system for landfill leachate treatment—challenges, opportunities, and recommendations. Geosyst Eng 19(6):337–345. https://doi.org/10.1080/12269328.2016.1188029

51. Kjeldsen P, Barlaz MA, Rooker AP, Baun A, Ledin A, Christensen TH (2002) Present and long-term composition of MSW landfill leachate: a review. Crit Rev Environ Sci Technol 32 (4):297–336
52. Contrera RC, da Cruz Silva KC, Morita DM, Domingues Rodrigues JA, Zaiat M, Schalch V (2014) First-order kinetics of landfill leachate treatment in a pilot-scale anaerobic sequence batch biofilm reactor. J Environ Manage 145:385–393. https://doi.org/10.1016/j.jenvman.2014.07.013
53. Cameron RD, Koch FA (1980) Trace metals and anaerobic digestion of leachate. J Water Pollut Control Fed 52(2):282–292
54. Luo J, Qian G, Liu J, Xu ZP (2015) Anaerobic methanogenesis of fresh leachate from municipal solid waste: a brief review on current progress. Renew Sustain Energy Rev 49:21–28. https://doi.org/10.1016/j.rser.2015.04.053
55. Dague RR, Habben CE, Pidaparti SR (1992) Initial studies on the anaerobic sequencing batch reactor. Water Sci Technol 26(9–11):2429–2432
56. Timur H, Özturk I (1999) Anaerobic sequencing batch reactor treatment of landfill leachate. Water Res 33(15):3225–3230. https://doi.org/10.1016/S0043-1354(99)00048-2
57. Kennedy KJ, Lentz EM (2000) Treatment of landfill leachate using sequencing batch and continuous flow upflow anaerobic sludge blanket (UASB) reactors. Water Res 34(14):3640–3656. https://doi.org/10.1016/S0043-1354(00)00114-7
58. Xie Z, Wang Z, Wang Q, Zhu C, Wu Z (2014) An anaerobic dynamic membrane bioreactor (AnDMBR) for landfill leachate treatment: performance and microbial community identification. Bioresour Technol 161:29–39. https://doi.org/10.1016/j.biortech.2014.03.014
59. Xiao Y, Yaohari H, De Araujo C, Sze CC, Stuckey DC (2017) Removal of selected pharmaceuticals in an anaerobic membrane bioreactor (AnMBR) with/without powdered activated carbon (PAC). Chem Eng J 321:335–345. https://doi.org/10.1016/j.cej.2017.03.118
60. Kawai M, Purwanti IF, Nagao N, Slamet A, Hermana J, Toda T (2012) Seasonal variation in chemical properties and degradability by anaerobic digestion of landfill leachate at Benowo in Surabaya, Indonesia. J Environ Manage 110:267–275. https://doi.org/10.1016/j.jenvman.2012.06.022
61. Sudibyo H, Shabrina ZL, Halim L, Budhijanto W (2017) Mathematical modelling and statistical approach to assess the performance of anaerobic fixed bed reactor for biogas production from Piyungan Sanitary Landfill leachate. Energy Procedia 105:256–262. https://doi.org/10.1016/j.egypro.2017.03.311
62. Ahmed FN, Lan CQ (2012) Treatment of landfill leachate using membrane bioreactors: a review. Desalination 287:41–54. https://doi.org/10.1016/j.desal.2011.12.012
63. Zolfaghari M, Droguia P, Brar SK, Buelna G, Dubé R (2016) Effect of bioavailability on the fate of hydrophobic organic compounds and metal in treatment of young landfill leachate by membrane bioreactor. Chemosphere 161:390–399. https://doi.org/10.1016/j.chemosphere.2016.07.021
64. Kizito S, Lv T, Wu S, Ajmal Z, Luo H, Dong R (2017) Treatment of anaerobic digested effluent in biochar-packed vertical flow constructed wetland columns: role of media and tidal operation. Sci Total Environ 592:197–205. https://doi.org/10.1016/j.scitotenv.2017.03.125
65. Kizito S, Luo H, Wu S, Ajmal Z, Lv T, Dong R (2017) Phosphate recovery from liquid fraction of anaerobic digestate using four slow pyrolyzed biochars: dynamics of adsorption, desorption and regeneration. J Environ Manage 201:260–267. https://doi.org/10.1016/j.jenvman.2017.06.057
66. Kizito S, Wu S, Kipkemoi Kirui W, Lei M, Lu Q, Bah H, Dong R (2015) Evaluation of slow pyrolyzed wood and rice husks biochar for adsorption of ammonium nitrogen from piggery manure anaerobic digestate slurry. Sci Total Environ 505:102–112. https://doi.org/10.1016/j.scitotenv.2014.09.096
67. Lo IMC (1996) Characteristics and treatment of leachates from domestic landfills. Environ Int 22(4):433–442. https://doi.org/10.1016/0160-4120(96)00031-1

68. Diamadopoulos E, Samaras P, Dabou X, Sakellaropoulos GP (1997) Combined treatment of landfill leachate and domestic sewage in a sequencing batch reactor. Water Sci Technol 36 (2):61–68. https://doi.org/10.1016/S0273-1223(97)00370-3
69. Mojiri A, Aziz HA, Zaman NQ, Aziz SQ, Zahed MA (2014) Powdered ZELIAC augmented sequencing batch reactors (SBR) process for co-treatment of landfill leachate and domestic wastewater. J Environ Manage 139:1–14. https://doi.org/10.1016/j.jenvman.2014.02.017
70. Mojiri A, Aziz HA, Zaman NQ, Aziz SQ, Zahed MA (2016) Metals removal from municipal landfill leachate and wastewater using adsorbents combined with biological method. Desalin Water Treat 57(6):2819–2833. https://doi.org/10.1080/19443994.2014.983180
71. Xie B, Xiong S, Liang S, Hu C, Zhang X, Lu J (2012) Performance and bacterial compositions of aged refuse reactors treating mature landfill leachate. Bioresour Technol 103 (1):71–77. https://doi.org/10.1016/j.biortech.2011.09.114
72. Nie F, Liu R, Li W, Liu Z (2016) Adsorption of COD and ammonia nitrogen in leachate from biochemical process on modified aged refuse adsorbent. Chin J Environ Eng 10 (6):2786–2792. https://doi.org/10.12030/j.cjee.201501115
73. Ismail S, Tawfik A (2016) Performance of passive aerated immobilized biomass reactor coupled with Fenton process for treatment of landfill leachate. Int Biodeterior Biodegradation 111:22–30. https://doi.org/10.1016/j.ibiod.2016.04.010
74. Sun H, Peng Y, Shi X (2015) Advanced treatment of landfill leachate using anaerobic–aerobic process: organic removal by simultaneous denitritation and methanogenesis and nitrogen removal via nitrite. Bioresour Technol 177:337–345. https://doi.org/10.1016/j.biortech.2014.10.152
75. Ogata Y, Ishigaki T, Ebie Y, Sutthasil N, Chiemchaisri C, Yamada M (2015) Water reduction by constructed wetlands treating waste landfill leachate in a tropical region. Waste Manage 44:164–171. https://doi.org/10.1016/j.wasman.2015.07.019
76. Huang W, Wang Z, Guo Q, Wang H, Zhou Y, Ng WJ (2016) Pilot-scale landfill with leachate recirculation for enhanced stabilization. Biochem Eng J 105(Part B):437–445. https://doi.org/10.1016/j.bej.2015.10.013
77. Benyoucef F, Makan A, El Ghmari A, Ouatmane A (2016) Optimized evaporation technique for leachate treatment: small scale implementation. J Environ Manage 170:131–135. https://doi.org/10.1016/j.jenvman.2015.12.004
78. J-h Im, H-j Woo, M-w Choi, K-b Han, C-w Kim (2001) Simultaneous organic and nitrogen removal from municipal landfill leachate using an anaerobic-aerobic system. Water Res 35 (10):2403–2410. https://doi.org/10.1016/S0043-1354(00)00519-4
79. Kettunen RH, Hoilijoki TH, Rintala JA (1996) Anaerobic and sequential anaerobic-aerobic treatments of municipal landfill leachate at low temperatures. Bioresour Technol 58(1):31–40. https://doi.org/10.1016/S0960-8524(96)00102-2
80. Ağdağ ON, Sponza DT (2005) Anaerobic/aerobic treatment of municipal landfill leachate in sequential two-stage up-flow anaerobic sludge blanket reactor (UASB)/completely stirred tank reactor (CSTR) systems. Process Biochem 40(2):895–902. https://doi.org/10.1016/j.procbio.2004.02.021
81. Wu L, Zhang L, Xu Y, Liang C, Kong H, Shi X, Peng Y (2016) Advanced nitrogen removal using bio-refractory organics as carbon source for biological treatment of landfill leachate. Sep Purif Technol 170:306–313. https://doi.org/10.1016/j.seppur.2016.06.033
82. Trabelsi I, Sellami I, Dhifallah T, Medhioub K, Bousselmi L, Ghrabi A (2009) Coupling of anoxic and aerobic biological treatment of landfill leachate. Desalination 246(1):506–513. https://doi.org/10.1016/j.desal.2008.04.059
83. Bakraouy H, Souabi S, Digua K, Dkhissi O, Sabar M, Fadil M (2017) Optimization of the treatment of an anaerobic pretreated landfill leachate by a coagulation–flocculation process using experimental design methodology. Process Saf Environ Prot 109:621–630. https://doi.org/10.1016/j.psep.2017.04.017
84. El-Gohary FA, Kamel G (2016) Characterization and biological treatment of pre-treated landfill leachate. Ecol Eng 94:268–274. https://doi.org/10.1016/j.ecoleng.2016.05.074

85. Oumar D, Patrick D, Gerardo B, Rino D, Ihsen BS (2016) Coupling biofiltration process and electrocoagulation using magnesium-based anode for the treatment of landfill leachate. J Environ Manage 181:477–483. https://doi.org/10.1016/j.jenvman.2016.06.067
86. Zolfaghari M, Jardak K, Drogui P, Brar SK, Buelna G, Dubé R (2016) Landfill leachate treatment by sequential membrane bioreactor and electro-oxidation processes. J Environ Manage 184:318–326. https://doi.org/10.1016/j.jenvman.2016.10.010
87. Zayen A, Schories G, Sayadi S (2016) Incorporation of an anaerobic digestion step in a multistage treatment system for sanitary landfill leachate. Waste Manage 53:32–39. https://doi.org/10.1016/j.wasman.2016.04.030
88. Torretta V, Ferronato N, Katsoyiannis I, Tolkou A, Airoldi M (2017) Novel and conventional technologies for landfill leachates treatment: a review. Sustainability 9(1):9
89. Bashir MJK, Aziz HA, Amr SSA, Sethupathi S, Ng CA, Lim JW (2015) The competency of various applied strategies in treating tropical municipal landfill leachate. Desalin Water Treat 54(9):2382–2395. https://doi.org/10.1080/19443994.2014.901189
90. de Morais JL, Zamora PP (2005) Use of advanced oxidation processes to improve the biodegradability of mature landfill leachates. J Hazard Mater 123(1):181–186. https://doi.org/10.1016/j.jhazmat.2005.03.041
91. Z-p Wang, Zhang Z, Y-j Lin, N-s Deng, Tao T, Zhuo K (2002) Landfill leachate treatment by a coagulation–photooxidation process. J Hazard Mater 95(1):153–159. https://doi.org/10.1016/S0304-3894(02)00116-4
92. Rivas FJ, Beltrán FJ, Gimeno O, Frades J, Carvalho F (2006) Adsorption of landfill leachates onto activated carbon. J Hazard Mater 131(1):170–178. https://doi.org/10.1016/j.jhazmat.2005.09.022
93. Turki N, Belhaj D, Jaabiri I, Ayadi H, Kallel M, Bouzid J (2013) Determination of organic compounds in landfill leachates treated by coagulation-flocculation and Fenton-adsorption. J Environ Sci Toxicol Food Technol 7(3):18–25
94. Maranon E, Castrillon L, Fernandez-Nava Y, Fernandez-Mendez A, Fernandez-Sanchez A (2009) Tertiary treatment of landfill leachates by adsorption. Waste Manage Res J Int Solid Wastes Public Cleansing Assoc ISWA 27(5):527–533. https://doi.org/10.1177/0734242x08096900
95. Suresh B, Thiruselvi D, Amudha T, Nilavunesan D, Sivanesan S (2016) Treatment of landfill leachate by using sequential batch reactor and sand bed filter followed by granular activated carbon (GAC). J Chem Pharm Sci 9(3):1468–1471
96. Nooten TV, Diels L, Bastiaens L (2008) Design of a multifunctional permeable reactive barrier for the treatment of landfill leachate contamination: laboratory column evaluation. Environ Sci Technol 42(23):8890–8895
97. Liyan S, Youcai Z, Weimin S, Ziyang L (2009) Hydrophobic organic chemicals (HOCs) removal from biologically treated landfill leachate by powder-activated carbon (PAC), granular-activated carbon (GAC) and biomimetic fat cell (BFC). J Hazard Mater 163(2):1084–1089. https://doi.org/10.1016/j.jhazmat.2008.07.075
98. Papastavrou C, Mantzavinos D, Diamadopoulos E (2009) A comparative treatment of stabilized landfill leachate: coagulation and activated carbon adsorption vs. electrochemical oxidation. Environ Technol 30(14):1547–1553
99. Chys M, Declerck W, Audenaert WTM, Van Hulle SWH (2015) UV/H_2O_2, O_3 and (photo-) Fenton as treatment prior to granular activated carbon filtration of biologically stabilized landfill leachate. J Chem Technol Biotechnol 90(3):525–533. https://doi.org/10.1002/jctb.4344
100. Lim CK, Seow TW, Neoh CH, Md Nor MH, Ibrahim Z, Ware I, Mat Sarip SH (2016) Treatment of landfill leachate using ASBR combined with zeolite adsorption technology. 3. Biotech 6(2):195. https://doi.org/10.1007/s13205-016-0513-8
101. Robinson T (2017) Removal of toxic metals during biological treatment of landfill leachates. Waste Manage 63:299–309. https://doi.org/10.1016/j.wasman.2016.12.032

102. Wang J, Liao Z, Ifthikar J, Shi L, Du Y, Zhu J, Xi S, Chen Z, Chen Z (2017) Treatment of refractory contaminants by sludge-derived biochar/persulfate system via both adsorption and advanced oxidation process. Chemosphere 185:754–763. https://doi.org/10.1016/j.chemosphere.2017.07.084
103. Rodríguez J, Castrillón L, Marañón E, Sastre H, Fernández E (2004) Removal of non-biodegradable organic matter from landfill leachates by adsorption. Water Res 38 (14):3297–3303. http://dx.doi.org/10.1016/j.watres.2004.04.032
104. Gao JL, Oloibiri V, Chys M, De Wandel S, Decostere B, Audenaert W, He YL, Van Hulle SWH (2015) Integration of autotrophic nitrogen removal, ozonation and activated carbon filtration for treatment of landfill leachate. Chem Eng J 275:281–287. https://doi.org/10.1016/j.cej.2015.04.012
105. Hur JM, Kim SH (2000) Combined adsorption and chemical precipitation process for pretreatment or post-treatment of landfill leachate. Korean J Chem Eng 17(4):433–437. https://doi.org/10.1007/bf02706856
106. Hur JM, Park JA, Son BS, Jang BG, Kim SH (2001) Mature landfill leachate treatment from an abandoned municipal waste disposal site. Korean J Chem Eng 18(2):233–239. https://doi.org/10.1007/bf02698465
107. Çeçen F, Erdinçler A, Kiliç E (2003) Effect of powdered activated carbon addition on sludge dewaterability and substrate removal in landfill leachate treatment. Adv Environ Res 7(3): 707–713. https://doi.org/10.1016/S1093-0191(02)00033-3

Current Progress on Removal of Recalcitrance Coloured Particles from Anaerobically Treated Effluent Using Coagulation–Flocculation

A. Y. Zahrim

Abstract The palm oil industry is the most important agro industries in Malaysia and most of the mills adopt anaerobic digestion as their primary treatment for palm oil mill effluent (POME). Due to the public concern, decolourisation of anaerobically treated POME (AnPOME) is becoming a great concern. Presence of recalcitrant-coloured particles hinders biological processes and coagulation–flocculation may able to remove these coloured particles. Several types of inorganic and polymers-based coagulant/flocculant aids for coagulation–flocculation of AnPOME have been reviewed. Researchers are currently interested in using natural coagulant and flocculant aids. Modification of the properties of natural coagulant and flocculant aids enhanced coagulation–flocculation performance. Modelling and optimization of the coagulation–flocculation process have also been reviewed. Chemical sludge has the potential for plant growth that can be evaluated through pot trials and phytotoxicity test.

Keywords Palm oil mill effluent · Anaerobic digestion · Decolourisation

1 Introduction

Palm oil is the world's most important oil crop that produces crude palm oil (CPO) and palm kernel oil (PKO), respectively. Total world production of CPO stands at about 38 million tonnes worth around US$20 billion [1]. The palm oil industry is the most important agro industry in Malaysia and an important contributor to Malaysia's economic growth. Other countries such as Indonesia, Thailand, Papua New Guinea, Colombia, Ivory Coast and Nigeria also plant oil palm [1].

A. Y. Zahrim (✉)
Chemical Engineering Programme, Faculty of Engineering, Universiti Malaysia Sabah, Jalan UMS, 88400 Kota Kinabalu, Sabah, Malaysia
e-mail: zahrim@ums.edu.my

N. Horan et al. (eds.), *Anaerobic Digestion Processes*,
Green Energy and Technology, https://doi.org/10.1007/978-981-10-8129-3_9

In Malaysia, the total production of crude palm oil (CPO) in 2012 alone was about 18.79 million tonnes [2]. However, the production of a large amount of crude palm oil (CPO) leads to enormous quantities of wastes, particularly palm oil mill effluent (POME). The minimum POME generated in 2012 is estimated to be 47 million tonnes [3]. Although there is no chemical addition during the production of CPO [4], the POME is a highly polluting wastewater that pollutes the environment if discharged directly into rivers (Table 1). POME is a colloidal mixture of water, oil and fine suspended components. The suspended components are mainly vegetative matter like cell walls, organelles, short fibres, water-soluble carbohydrates ranging from hemicelluloses to simple sugars (glucose, reducing sugars and pectin), nitrogenous compounds (from proteins to amino acids), free organic acids, lipids, as well as the assembly of minor organic and mineral constituents. The suspended solids in POME slurry are mainly cellulose matter mixed with small portions of residue oil [5].

Due to the rapid development of palm oil industry worldwide, the risks of pollution generated from the industry have been growing. For example, the number of palm oil mills in Malaysia continued to increase rapidly from 334 mills in 1999 to 454 mills in 2017 and Abdullah et al. [8] estimated that the current POME volumes will be increasing from 60 million tonnes to 70–110 million tonnes by 2020.

In order to control the pollution from palm oil mills in the country, regulatory control over discharges from palm oil mills is instituted through Environmental Quality (Prescribed Premises) (Crude Palm Oil) Regulations, 1977 promulgated under the Environmental Quality Act, 1974 and enforced by the Department of Environmental (DOE). The palm oil mills are required to adhere to prescribed regulations, which include laws governing the discharge of mill effluent into watercourses and to land. Moreover, the requirement for the BOD concentration of industrial effluents to be discharged to watercourse has been made tighter recently

Table 1 Characteristics of raw POME and AnPOME

Parameter[a]	POME [6]	AnPOME [7]
pH	3.4–5.2	7.2–8.3
Biochemical oxygen demand	10,250–43,750	440–1355
Chemical oxygen demand	15,000–100,000	1003–13,532
Total solids	11,500–79,000	NS
Suspended solids	5000–54,000	290–12,750
Volatile solids	9000–72,000	NS
Oil and grease	130–18,000	NS
Ammoniacal nitrogen	29–312	45–100
total nitrogen	180–1400	26–310

[a]Units in mg/L except pH, NS = not stated

by DOE where the prevailing national regulation of 100 mg/L BOD has now been reduced to 20 mg/L for mills [3, 8]. The regulations also outlined the effluent discharge standard to comply with the colour discharge of 100 ADMI [8].

Anaerobic digestion of POME is widely accepted by the managers due to its low operating cost and biogas recovery, which reduces the carbon footprint of palm oil production [9]. Although there is significant organic matter reduction during anaerobic digestion treatment, the colour of effluent (AnPOME) turns dark brown (Fig. 1) [10] and it is also contains bioflocs, anaerobic microorganisms and macrofibrils [11]. The ineffectiveness of POME anaerobic digestion is due to insufficient bacterial activity stunted by irregular effluent flows, volatile POME components, complex POME compositions, rainwater dilution, remaining suspended solids, crystal-like struvite formation of POME organic fouling, as well as unfiltered solubilised plant proteins and sugars [9]. Figure 2a shows particles from anaerobic ponds that consist of loose strand fibres. It is believed that after it is undergone degradation, the loose strands fibres is disappeared as shown in Fig. 2b.

Colour is the first contaminant to be recognised by the public and hence the appearance from the anaerobically treated POME (AnPOME) becomes a great concern. Discharge of coloured AnPOME imparts colour to receiving waters and thus inhibits the growth of marine organisms by reducing the penetration of sunlight, with a consequent reduction in photosynthetic activity. The coloured compounds may chelate with metal ions and thus become directly toxic to aquatic biota [12]. Besides, the humic substances will react with chlorine in drinking water treatment and produce carcinogenic by-products such as trihalomethanes [13]. In addition, substances derived from lignin in POME can possibly inhibit embryonic development in marine organisms [14]. Several studies (e.g. Jakobsen et al. [15]; Fathahi [16] reported the occurrence of water pollution which is caused by improper treatment of palm oil mill effluent (POME). The colour of the effluent

Fig. 1 **a** Raw POME **b** anaerobically treated POME [22]

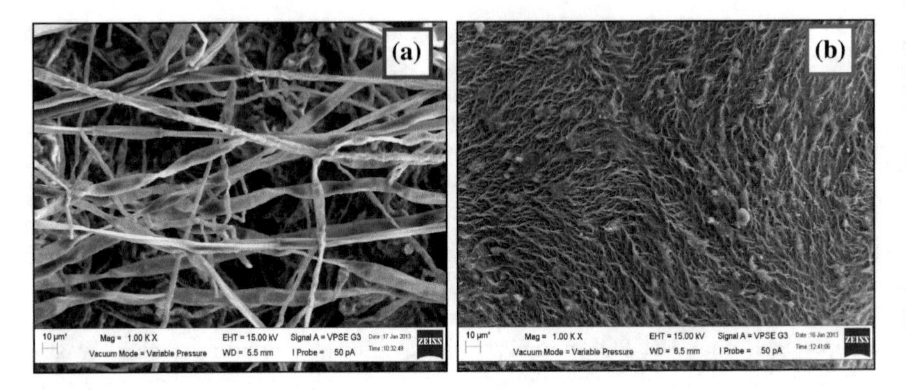

Fig. 2 SEM images of particles from **a** anaerobic pond and **b** aerobic pond, at 1000× magnification

might be contributed by the residual lignin [17], tannin, humic and fulvic acid-like substance [18, 19], lipids and fatty acids [20] as well as anaerobic fermentation by-product, e.g. melanoidin [4, 21].

Treating coloured effluent from industries has become a real challenge in the recent years. The objective of this review is to discuss the current progress on the treatment of AnPOME using coagulation–flocculation. The coagulation–flocculation process is regarded as one of the most important and widely used treatment processes for industrial wastewater [23] and raw water [24] due to its simplicity and effectiveness. Utilising coagulation–flocculation process for AnPOME decolourisation could enhance recovery and recycling of water for the palm oil mill [25].

2 Anaerobic Digestion of Palm Oil Mill Effluent

Anaerobic digestion (AD) is one of the oldest known processes utilised for the metabolism of organic wastes. The basis of this treatment method is the evolution of methane via degradation in the absence of oxygen. The numerous advantages of anaerobic digestion include the low operating costs as minimal chemicals required, pathogen removal, higher loading rates are possible and the formation of biogas from the metabolism of more than 90% of organic material [26].

Generally, the anaerobic digestion mechanism consists of a number of “stages”: (1) hydrolysis, (2) fermentation (acidogenesis) and (3) methanogenesis [27]. During hydrolysis, two mechanisms may take place: (1) the particulate material (i.e. plant cell debris and less than 50% of total pollutant level [28]) is converted to the soluble compound [27] and/or (2) the complex material (carbohydrate, lipids and protein) is converted to simple compounds (sugar, amino acids, etc.) [29].

However, anaerobic digestion is not efficient for the removal of coloured compounds/phenolics due to their inhibitory effects on anaerobic microorganisms [30, 31].

3 Coloured Compounds

Lignin, the main plant component, is a heterogeneous aromatic polymer interspersed with hemicellulose and occurs surrounding microfibrils. Lignin (density: 1.3–1.4 g cm^{-3} and brown in colour) contains P-hydroxy-phenyl, syringyl and guaiacyl units [12]. The lignin content in POME is around 1700–7890 mg/L [17, 32].

Another important colourant that could be present in AnPOME might be tannin [18]. Tannins are complex dark-coloured non-crystalline substances composed of polyhydroxy phenolic (aromatic hydroxyl) compounds, related to catechol, glycosides or pyrogallol, which vary in composition. Tannins extracted from wood, bark and leaves are used extensively in the preservation of animal skins [12].

During the fermentation stage, the amino acids, sugars and fatty acids are degraded to several compounds, i.e. lactate, propionate, acetate, formate, etc. [27]. Besides that, natural condensation between sugars (carbonyl groups) and amino acids or proteins (free amino groups) through the Maillard reaction could produce another colourant, i.e. melanoidin [33]. It has been reported that the wastewater from distilleries and fermentation industries also contain melanoidins [33]. Presently, there is no report of Maillard reaction products in AnPOME. Finally, during anaerobic digestion the methanogenic substrates are converted to methane and carbon dioxide [27]. Due to its structural complexity, dark colour and offensive odour, it poses a serious threat to soil and aquatic ecosystem [33]. The offensive odour might be due to the presence of volatile fatty acids such as butyric and valeric acid [34].

4 Coagulation–Flocculation Process

The attractive forces between the particles are considerably less than the repelling forces of electric charge. Under these conditions, Brownian motion keeps the particles in stable suspension. Due to this, the particles will not settle out on standing or may settle by taking a very long time. Coagulation–flocculation in effluent treatment involves the addition of substances to alter the physical state of colloidal and suspended particles. Coagulation–flocculation expedites particle removal by sedimentation. Coagulation tends to overcome the factors that promote effluent particle stability and form agglomerates or flocs. In water and effluent treatment, the coagulation–flocculation process is extremely vital [35]. Coagulation as a pretreatment is regarded as the most successful pretreatment for water treatment

[36]. Besides that, coagulation–flocculation of coloured effluents has been used for many years, either as a main or pre-treatment option, due to its low capital cost [37]. Several factors are affecting the coagulation–flocculation process, i.e. types of coagulant/flocculant aids, coagulant/flocculant aids dosage, mixing rate and time, pH and temperature.

Coagulants can be categorised as [38]: (1) hydrolysing metallic salts, e.g. ferric chloride, aluminium sulphate (alum), etc., (2) pre-hydrolysing metallic salts, e.g. polyaluminium chloride, polyferrous sulphate, etc., (3) synthetic cationic polymers, e.g. polydiallyldimethyl ammonium chloride, polyacrylamide, polyamine, etc. (4) plant-based natural coagulants, e.g. guar gum, potato starch, *Moringa oleifera* seeds, etc. (5) animal-based natural coagulants, e.g. chitosan, fish scale, etc., and (6) microorganism-based natural coagulants, e.g. xanthan gum.

Since coloured particles in AnPOME are negatively charged [31, 39]; cationic coagulants such as aluminium sulphate [11, 40, 41], aluminium chlorohydrate [42], ferric chloride [39, 43], polyaluminium chloride [44, 45], calcium lactate [31], chitosan [46] as well as polydiallyldimethyl ammonium chloride [42] have been investigated for coagulation purpose.

Flocculation is the process whereby destabilised particles, or particles formed as a consequence of destabilisation, are induced to come together, make contact, and thereby form large(r) agglomerates [35]. Flocculation can happen through the addition of coagulant alone (called as primary coagulant) or flocculant aids. Normally, anionic flocculant aids are being used for this purpose: anionic polyacrylamide [39, 43], rice starch [47], etc. Anionic flocculant aids are useful when the destabilised flocs have excess positive charge surface and bridging between flocs has occurred. In some cases, cationic polyacrylamide might be used as flocculant aids [11, 31] if the destabilised flocs still contain an excess of negative charge surface. It should be noted that precipitation (chemical state alteration) might occurs simultaneously during coagulation–flocculation process [48, 49]. The best polymer should be selected to ensure the highest performance of coagulation as well as to reduce the chemical cost [50].

4.1 Inorganic and Polymers-Based Coagulant/Flocculant Aids

In Malaysia, a coagulation–flocculation (CF) treatment method has been investigated for the treatment of raw POME as well anaerobically treated POME (AnPOME). While in raw POME the treatment focuses on suspended solids removal, coagulation–flocculation of AnPOME is aiming for removal of soluble solids including recalcitrance coloured particles.

Ho and Tan [11] studied AnPOME treatment through coagulation–flocculation (CF) using aluminium sulphate-cationic polyacrylamide, dissolved air flotation (DAF) and CF-DAF to reduce conventional treatment time and area. Despite the fact that both methods were able to achieve a 97% removal of the suspended solids of the AnPOME, the removal of the soluble solid is very difficult. The authors stated that the total solid removal for CF, DAF and CF-DAF are 56, 59 and 63%, respectively [11]. Despite advancement in polymer synthesising and purification resulting in the development of vast types of polymers, the performance of polymers towards soluble solid is still unchanged. Malakahmad et al. [41] investigated the further treatment of AnPOME using alum combined with cationic polymers. Results indicate coagulation process under optimum conditions (pH = 6, alum dosage = 1800 mg/L, rapid mixing = 5 min, and slow mixing = 20 min) reduces the COD, TSS and turbidity by 59, 80 and 86, respectively. The polymers caused further reduction of TSS (85–88%) and turbidity (97–98%) but not the soluble COD [41].

Polyaluminium chloride (PAC) could be effective at neutral pH of initial wastewater. Poh et al. [44] investigated polyaluminium chloride (PAC) as a post-treatment method for POME. Hybrid PAC (2 g/L) with micro-bubbles system able to reduce COD about 93% at pH 7.05 and final treated effluent pH is within the regulatory requirements, i.e. 6.28 [44]. In another study, Othman et al. [45] showed that a hybrid process, i.e. adsorption using activated carbon with coagulation using polyaluminium chloride (PAC) resulted in the final COD and SS of 10 and 2 mg/L, respectively, which is better than river water quality [45].

Jami et al. [43] applied anionic polymer as a flocculant aid and compared the use of coagulants ferric chloride and aluminium sulphate to reduce turbidity. The result of the coagulation process showed that ferric chloride gave a better reduction of turbidity at a dosage of 100 mg/L, pH of 8 and with polymer dose of 100 mg/L than alum. Similar findings were also observed by Othman et al. [45], i.e. $FeCl_3$ showed highest COD and suspended solid (SS) removal compared to alum and PAC [45].

Zinatizadeh et al. [51] evaluated various biodegradable polymers, i.e. three cationic polyacrylamides (C-PAM; as coagulant) and three anionic polyacrylamides (A-PAM; as flocculant) with different molecular weights and charge densities, for the treatment of POME. The combination of a C-PAM (Chemfloc1515C) with medium molecular weight and charge density and an A-PAM (Chemfloc 430A) with high molecular weight and charge density at doses of 300 and 50 mg/dm^3 showed the best total suspended solids (TSS) and chemical oxygen demand (COD) removal (96.4 and 70.9%, respectively). The optimal condition was found at pH 5, rapid mixing at 150 rpm for 1 min, and slow mixing at 40 rpm for 30 s. As a conclusion, physiochemical pretreatment using biodegradable coagulants was a promising alternative to effectively separate TSS (96.4%) with high water recovery (76%) [51].

The most important part of sludge treatment prior to disposal is the reduction of the sludge volume by solid–liquid separation. However, the presence of organic components, mainly bacterial cells and extracellular polymeric substances (EPS), and colloidal and supracolloidal range particles in the sludge makes it difficult to dewater even at high pressures, i.e. 0.6 MPa [52]. Zinatizadeh et al. [51] evaluated various biodegradable polymers, i.e. three cationic polyacrylamides (C-PAM; as coagulant) and three anionic polyacrylamides (A-PAM; as flocculant) with different molecular weights and charge densities; for the treatment of POME. The combination of a C-PAM (Chemfloc1515C) with medium molecular weight and charge density and an A-PAM (Chemfloc 430A) with high molecular weight and charge density at doses of 300 and 50 mg/dm^3 showed the best total suspended solids (TSS) and chemical oxygen demand (COD) removal (96.4 and 70.9%, respectively). The optimal condition was found at pH 5, rapid mixing at 150 rpm for 1 min, and slow mixing at 40 rpm for 30 s. As a conclusion, the physiochemical pretreatment using biodegradable coagulants was a promising alternative to effectively separate TSS (96.4%) with high water recovery (76%) [51].

4.2 Natural Coagulant and Flocculant Aids

Due to the increasing awareness of the toxicity of inorganic coagulants, several investigations have been carried out to replace/minimise inorganic coagulants with non-toxic and biodegradable coagulants/flocculant aids. Chitosan is a natural organic polyelectrolyte of high molecular weight and charge density; obtained from deacetylation of chitin. Abu-Hasan and Puteh [53] explored the potential and effectiveness of applying chitosan as a primary coagulant and flocculant aids, in comparison with aluminium sulphate (alum) for pre-treatment of palm oil mill effluent (POME). Chitosan showed better parameter reductions with much lower dosage compared to alum. At pH 6, the optimum chitosan dosage of 400 mg/L was able to reduce turbidity, TSS and COD levels by 99.90, 99.15 and 60.73% respectively. At this pH, the coagulation of POME by chitosan was brought about by the combination of charge neutralisation and a polymer bridging mechanism. It can be suggested that polymer bridging by chitosan is more dominant than alum and the dosage of alum can also be reduced [53]. Parthasarathy et al. [46] studied treatment of AnPOME through the following 4 strategies: coagulation by chitosan, the addition of ferrous sulphate ($FeSO_4$), chitosan with hydrogen peroxide (H_2O_2) and chitosan with Fenton oxidation. Coagulation only by using chitosan (2500 mg/L) achieved the maximum COD and TSS removal of 70.22 and 85.59%, respectively. In conclusion, they reported that chitosan with H_2O_2 proved to be the most promising alternative for POME treatment compared to chitosan with Fenton oxidation [46].

Teh et al. [47] investigated the use of various starches, i.e. rice starch, wheat starch, corn starch and potato starch, to replace alum for POME treatment. Rice starch was found to be the best starch based on the removal of total suspended solids (TSS). The use of rice starch alone at room temperature enabled the removal of TSS up to 84.1% using the recommended values of dosage, initial pH, settling time and slow stirring speed at 2 g/L, pH 3, 5 min and 10 rpm, respectively. Higher TSS removal of 88.4% could still be achieved at a lower dosage of rice starch (0.55 g/L) only when rice starch was used together with 0.2 g/L of alum during the treatment of POME [47]. Although alum can efficiently reduce the pollutant, the final treated water is acidic and need to be neutralised before it can be discharged. In addition, the process is only efficient at the acidic condition and as a result of this, the initial effluent needs to be acidified first before undergoing coagulation–flocculation, since the original pH for AnPOME is between 7.2 and 8.3 [7]. Coagulation–flocculation of AnPOME without changing the initial pH using calcium lactate was investigated by Zahrim et al. [31]. The best polymer order was identified based on an overall removal performance. The best polymer can be arranged as QF23912 (58%) > QF25610 (57%) > AN1500 (51%) > QF24807 (50%) > AN1800 (47%). All tested polymers have similarity in removing NH3-N [31].

Cassia obtusifolia, is a weed that is abundant in Asia. Earlier investigations by Shak and Wu [54] showed that generally, *C. obtusifolia* seed gum performed better than alum in removing TSS and COD from the POME. It is interesting to note that wastewater temperature had a negligible effect on the treatment efficiency when *C. obtusifolia* seed gum was used compared to alum, which was more sensitive to temperature change. Optimised treatment conditions when using *C. obtusifolia* seed gum for the treatment of POME (7500 mg/L) were determined and required a natural coagulant dosage of 1.0 g/L, initial pH of 3 and a settling time of 45 min [54]. Birima et al. [55] investigated the effectiveness of salt extracted peanut seeds after oil extraction as a coagulant. The active coagulation component was extracted using three different concentrations of sodium chloride (NaCl), namely 0, 1 and 2 mol/L. The authors reported that the higher NaCl concentration resulted in a low optimum dosage of peanut seeds and higher removal of turbidity, TSS and COD. Peanut seeds extracted with 2 mol/L reduced TSS to 1218 mg/l (94.7% removal). On the other hand, peanut seeds extracted with distilled water reduced TSS to 2175 mg/L (90% removal) [55].

Shak and Wu [56] developed a cationic plant-based seed gum derived from *C. obtusifolia* for treatment of POME. Quaternized *C. obtusifolia* seed gum (seed gum-CHPTAC) was obtained through seed gum etherification with *N*-(3-chloro-2-hydroxypropyl) trimethyl ammonium chloride (CHPTAC). The influence of cationic monomer concentration, catalyst concentration, reaction temperature and reaction time were studied for the synthesis based on total suspended solids (TSS) and chemical oxygen demand (COD) removals from the POME. It was concluded that the changes in properties led to its superior effectiveness as compared to its natural seed gum form in the treatment of POME [56]. Coagulation–flocculation using Organo-floc, a vegetable-based cationic organic polymer shows the capability to treat AnPOME as great as alum does. Organo-floc can remove almost 71% of solid from the wastewater by comparing to alum which only can remove at 65% of solid but for COD removal alum show 50% removal higher efficiencies of removal compared to Organo-floc [25].

4.3 Modelling and Optimization

Optimisation of the coagulation process can be determined using central composite design (CCD) and response surface methodology (RSM) [57]. By employing CCD-RSM, Malakahmad and Chuan [40] have successfully modelled the interaction between several parameters, i.e. pH, alum dosage and mixing rate, with AnPOME. Results show the regression, linear, interaction and quadratic terms are significant and the model is considered to be adequate in terms of reproducibility. After operating of the coagulation process under optimum condition (pH = 6.4, alum dosage = 2124 mg/L, and slow mixing = 20 min) the chemical oxygen demand (COD) reduced by 59% [40].

Zahrim et al. [58] investigated a model solution containing lignin using a single mixing tank system approach with polydiallyldimethyl ammonium chloride (polyDADMAC) as a coagulant. Calcium lactate performed better than magnesium hydroxide and anionic polyacrylamide as flocculant aids. The coagulation/flocculation with polyDADMAC-calcium lactate removed lignin through a complex mechanism: the adsorptive-charge neutralisation-precipitation-bridging mechanism. Response surface methodology (RSM) study indicated that strong interaction in the coagulation/flocculation of lignin occurred between the initial pH-polyDADMAC dosage, initial pH-calcium lactate dosage and polyDADMAC-calcium lactate dosage. The highest lignin removal achieved was between 50 and 68% [58].

Recently, Tamrin and Zahrim [59] determined the best flocculant aids in the coagulation–flocculation process of AnPOME by considering all output responses, namely lignin–tannin, low molecular mass-coloured compounds (LMMCC), chemical oxygen demand (COD), ammonia nitrogen (NH_3-N), pH, and conductivity. Here, multiple-objective optimisation on the basis of ratio analysis (MOORA) is employed to discretely measure multiple response characteristics of five different types of flocculant aids as a function of assessment value. This study highlights the simplicity of MOORA approach in handling various input and output parameters [59].

4.4 Sludge Utilisation as Soil Conditioner

Mohd Tadza et al. [60] evaluated the sludge obtained from coagulation–flocculation of POME using chitosan-based coagulant by setting a set of pot trial tests using *Scindapsus aureus*. The comparison was made using commercially available fertiliser. It was found that after coagulation–flocculation, macronutrients such as nitrogen (N) and phosphorus (P) initially contained within POME were removed except for potassium (K). Nevertheless, pot trial test results indicated that *S. aureus* grows better using chitosan-based sludge as compared to commercially available fertiliser [60]. The performance of several chemical coagulants including ferric

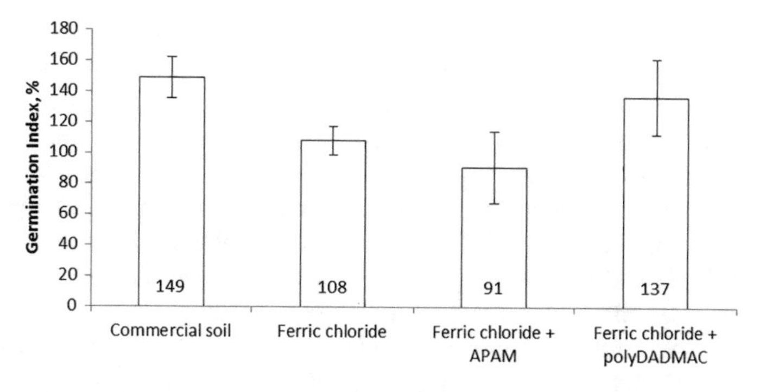

Fig. 3 The germination index (GI%) of cabbage seeds in the water-soluble extracts at different types of dry solids sample (50% commercial soil +50% dry solids) [39]

chloride, calcium lactate, magnesium hydroxide, aluminium chlorohydrate, and polydiallyldimethylammonium chloride (polyDADMAC) were investigated in removing colour from AnPOME The results show that ferric chloride as a sole coagulant can achieve a colour removal of more than 80% without the need for pH adjustment, which indicates the effectiveness of the coagulant to treat AnPOME. However, ferric chloride-anionic polyacrylamide (A-PAM) shows better performance than ferric chloride-polyDADMAC in terms of colour removal, pH, with shorter sedimentation time and the sludge has potential to be reused for land application [39]. It is believed that accumulation of higher polymer, i.e. A-PAM in the sludge contributes to the lower GI (Fig. 3) in the same study. Effects of Cu and cationic polymer flocculants on hydroponically cultured plants were studied by Kuboi and Fujii [61]. They suggest that: (1) chlorosis caused by cationic polymer flocculants is related to their Cu-holding capacity and (2) the growth reduction is firstly caused by the adhesion of the polymers to the roots and secondly by the toxicity of Cu which accumulates in the roots [61].

In another study, Kuboi and Fujii [62] studied 44 commercial products of synthetic polymer flocculants for phytotoxicity test using the turnip root assay. Of all the flocculants tested, only the cationic ones inhibited root elongation of turnip (*Brassica rapa* L.) at a concentration of less than 25 mg/L. Studies in the agricultural field by using A-PAM showed that the application of A-PAM to the soil may improve the soil permeability, stabilising the soil structure, minimising dispersion, and encourage aggregate formation to enhance pore continuity [63].

5 Conclusion and Future Works

In this review, several important studies on current coagulation–flocculation were highlighted. Although the application of polymers for AnPOME treatment has been initiated since the 1980s and many new polymers have synthesised and commercialised since then, it seems that the performance of coagulation–flocculation towards soluble particles is still unsatisfied. Therefore, many investigations focused on a hybrid system such as using H_2O_2, Fenton oxidation, etc.

Other than that several innovations have been introduced such as single tank [58] and utilising hybrid composite system [64] that could be a very interesting area of research in the future. One major hurdle for the coagulation–flocculation process is the sensitivity of the coagulant/flocculant aids to pollutant fluctuation [58]. In addition lack of systematic studies on utilisation of chemical sludge generated from flocculation. The effect of the chemical sludge on plant growth is also worth to investigate. Other than many possibilities of reusability of the sludge generated, e.g. lignin as fuel or application of biodrying for dewatering chemical sludge should be further explored.

References

1. Soh AC, Wong CK, Ho YW, Choong CW (eds) (2009) Oil palm. Oil crops, handbook of plant breeding 4. Springer
2. Annual and Forecast of Crude Palm Oil Production (Tonnes) 2011 & 2012 (2013). http://bepi.mpob.gov.my/index.php/statistics/production/71-production-2012/296-annual-forecast-production-of-crude-palm-oil-2011-2012.html
3. Chin MJ, Poh PE, Tey BT, Chan ES, Chin KL (2013) Biogas from palm oil mill effluent (POME): opportunities and challenges from Malaysia's perspective. Renew Sustain Energy Rev 26:717–726. https://doi.org/10.1016/j.rser.2013.06.008
4. Zahrim AY, Rachel FM, Menaka S, Su SY, Melvin F, Chan ES (2009) Decolourisation of anaerobic palm oil mill effluent via activated sludge-granular activated carbon. World Appl Sci J 5:126–129
5. Liew WL, Kassim MA, Muda K, Loh SK, Affam AC (2015) Conventional methods and emerging wastewater polishing technologies for palm oil mill effluent treatment: a review. J Environ Manage 149:222–235. https://doi.org/10.1016/j.jenvman.2014.10.016
6. MPOB (2013) Malaysian palm oil board. http://www.mpob.gov.my/palm-info/environment/520-achievements#Environmental; http://www.mpob.gov.my/palm-info/environment/520-achievements. Accessed 1 July 2013
7. Zahrim AY (2014) Palm oil mill biogas producing process effluent treatment: a short review. J Appl Sci 14(23):3149–3155
8. Abdullah N, Yuzir A, Curtis TP, Yahya A, Ujang Z (2013) Characterization of aerobic granular sludge treating high strength agro-based wastewater at different volumetric loadings. Biores Technol 127:181–187. https://doi.org/10.1016/j.biortech.2012.09.047
9. Loh SK, Nasrin AB, Mohamad Azri S, Nurul Adela B, Muzzammil N, Daryl Jay T, Stasha Eleanor RA, Lim WS, Choo YM, Kaltschmitt M (2017) First report on Malaysia's experiences and development in biogas capture and utilization from palm oil mill effluent under the economic transformation programme: current and future perspectives. Renew Sustain Energy Rev 74:1257–1274. https://doi.org/10.1016/j.rser.2017.02.066
10. Zahrim AY, Fansuri MB, Nurmin B, Rosalam S (2012) A review on the decolourisation of anaerobically treated palm oil mill effluent (AnPOME). In: Bono A, Sipaut CS (eds) Proceedings of 26th symposium of Malaysian chemical engineers, Kota Kinabalu, Sabah, Malaysia
11. Ho CC, Tan YK (1989) Comparison of chemical flocculation and dissolved air flotation of anaerobically treated palm oil mill effluent. Water Res 23(4):395–400

12. Mohan SV, Karthikeyan J (1997) Removal of lignin and tannin colour from aqueous solution by adsorption onto activated charcoal. Environ Pollut 97(1–2):183–187. https://doi.org/10.1016/S0269-7491(97)00025-0
13. Vukovic M, Domanovac T, Briki F (2008) Removal of humic substances by biosorption. J Environ Sci 20(12):1423–1428. https://doi.org/10.1016/S1001-0742(08)62543-7
14. Pillai MC, Blethrow H, Higashi RM, Vines CA, Cherr GN (1997) Inhibition of the sea urchin sperm acrosome reaction by a lignin-derived macromolecule. Aquat Toxicol 37(2–3):139–156. https://doi.org/10.1016/s0166-445x(96)00821-1
15. Jakobsen F, Hartstein N, Frachisse J, Golingi T (2007) Sabah shoreline management plan (Borneo, Malaysia): ecosystems and pollution. Ocean Coast Manag 50(1–2):84–102
16. Fathahi TKT (2010) Water quality and sources of pollution of the Sg Kinabatangan basin. In: Seminar and workshop POMET^3 Sabah, Malaysia
17. Poh PE, Yong W-J, Chong MF (2010) Palm oil mill effluent (POME) characteristic in high crop season and the applicability of high-rate anaerobic bioreactors for the treatment of POME. Ind Eng Chem Res 49(22):11732–11740. https://doi.org/10.1021/ie101486w
18. Edem DO (2002) Palm oil: biochemical, physiological, nutritional, hematological and toxicological aspects: a review. Plant Foods Hum Nutr (Formerly Qualitas Plantarum) 57 (3):319–341. https://doi.org/10.1023/a:1021828132707
19. Kongnoo A, Suksaroj T, Intharapat P, Promtong T, Suksaroj C (2012) Decolorization and organic removal from palm oil mill effluent by Fenton's Process. Environ Eng Sci 29(9):855–859. https://doi.org/10.1089/ees.2011.0181
20. Oswal N, Sarma PM, Zinjarde SS, Pant A (2002) Palm oil mill effluent treatment by a tropical marine yeast. Bioresour Technol 85(1):35–37
21. Bunrung S, Prasertsan S, Prasertsan P (2011) decolourisation of biogas effluent of palm oil mill using palm ash. In: Paper presented at the TIChE international conference 2011, Hatyai, Songkhla, Thailand, 10–11 Nov 2011
22. Yaser A, Nurmin B, Rosalam S (2013) Coagulation/flocculation of anaerobically treated palm oil mill effluent (AnPOME): a review. In: Developments in sustainable chemical and bioprocess technology. Springer, pp 3–9
23. Teh CY, Budiman PM, Shak KPY, Wu TY (2016) Recent advancement of coagulation-flocculation and its application in wastewater treatment. Ind Eng Chem Res 55 (16):4363–4389. https://doi.org/10.1021/acs.iecr.5b04703
24. Teh CY, Wu TY (2014) The potential use of natural coagulants and flocculants in the treatment of urban waters. Chem Eng Trans 39:1603–1608
25. Tajuddin HA, Abdullah LC, Idris A, Choong TSY (2015) Effluent quality of anaerobic palm oil mill effluent (POME) wastewater using organic coagulant. Int J Sci Res (IJSR) 4(5): 667–677
26. Ramsamy D, Rakgotho T, Naidoo V, Buckley CA (2002) Anaerobic co-digestion of high strength/toxic organic liquid effluents in a continuously stirred reactor: start-up. In: Biennial conference of the water Institute of Southern Africa (WISA) Durban, South Africa, 19–23 May 2002. Water Research Commission (WRC)
27. Metcalf E (2004) Wastewater engineering—treatment and reuse, 4th edn. McGraw-Hill Companies, New York
28. Wu TY, Mohammad AW, Jahim JM, Anuar N (2010) Pollution control technologies for the treatment of palm oil mill effluent (POME) through end-of-pipe processes. J Environ Manage 91(7):1467–1490
29. Poh PE, Chong MF (2009) Development of anaerobic digestion methods for palm oil mill effluent (POME) treatment. Biores Technol 100(1):1–9
30. Kietkwanboot A, Tran HTM, Suttinun O (2015) Simultaneous dephenolization and decolorization of treated palm oil mill effluent by oil palm fiber-immobilized trametes Hirsuta strain AK 04. Water Air Soil Pollut 226(10):345. https://doi.org/10.1007/s11270-015-2599-8
31. Zahrim AY, Nasimah A, Hilal N (2014) Pollutants analysis during conventional palm oil mill effluent (POME) ponding system and decolourisation of anaerobically treated POME via

calcium lactate-polyacrylamide. J Water Process Eng 4(4):159–165. https://doi.org/10.1016/j.jwpe.2014.09.005
32. Hii K-L, Yeap S-P, Mashitah MD (2012) Cellulase production from palm oil mill effluent in Malaysia: economical and technical perspectives. Eng Life Sci 12(1):7–28. https://doi.org/10.1002/elsc.201000228
33. Chandra R, Bharagava RN, Rai V (2008) Melanoidins as major colourant in sugarcane molasses based distillery effluent and its degradation. Biores Technol 99(11):4648–4660
34. Boopathy MA, Senthilkumar SNS (2014) Media optimization for the decolorization of distillery spent washby biological treatment using Pseudomonas fluorescence. Int J Innov Eng Technol 4(1):8–15
35. Bratby J (2006) Coagulation and flocculation in water and wastewater treatment. IWA Publishing
36. Leiknes T (2009) The effect of coupling coagulation and flocculation with membrane filtration in water treatment: a review. J Environ Sci-China 21(1):8–12
37. Anjaneyulu Y, Sreedhara Chary N, Samuel Suman Raj D (2005) Decolourization of industrial effluents—available methods and emerging technologies—a review. Rev Environ Sci Bio/Technol 4(4):245–273. https://doi.org/10.1007/s11157-005-1246-z
38. Verma AK, Dash RR, Bhunia P (2012) A review on chemical coagulation/flocculation technologies for removal of colour from textile wastewaters. J Environ Manage 93(1):154–168. https://doi.org/10.1016/j.jenvman.2011.09.012
39. Zahrim AY, Dexter ZD, Joseph CG, Hilal N (2017) Effective coagulation-flocculation treatment of highly polluted palm oil mill biogas plant wastewater using dual coagulants: decolourisation, kinetics and phytotoxicity studies. J Water Process Eng 16:258–269. https://doi.org/10.1016/j.jwpe.2017.02.005
40. Malakahmad A, Chuan SY (2013) Application of response surface methodology to optimize coagulation-flocculation treatment of anaerobically digested palm oil mill effluent using alum. Desal Water Treat 51(34–36):6729-6735. https://doi.org/10.1080/19443994.2013.791778
41. Malakahmad A, Chuan SY, Eisakhani M (2014) Post-treatment of anaerobically digested palm oil mill effluent by polymeric flocculant-assisted coagulation. Appl Mech Mat 567. doi:10.4028/www.scientific.net/AMM.567.116
42. Zahrim AY, Dexter ZD (2016) Decolourisation of palm oil mill biogas plant wastewater using poly-diallyldimethyl ammonium chloride (polyDADMAC) and other chemical coagulants. IOP Conference Series: Earth Environ Sci. https://doi.org/10.1088/1755-1315/36/1/012025
43. Jami MS, Muyibi SA, Oseni MI (2012) Comparative study of the use of coagulants in biologically treated palm oil mill effluent (POME). Adv Nat Appl Sci 6(5):646–650
44. Poh PE, Ong WYJ, Lau EV, Chong MN (2014) Investigation on micro-bubble flotation and coagulation for the treatment of anaerobically treated palm oil mill effluent (POME). J Environ Chem Eng 2(2):1174–1181. https://doi.org/10.1016/j.jece.2014.04.018
45. Othman MR, Hassan MA, Shirai Y, Baharuddin AS, Ali AAM, Idris J (2014) Treatment of effluents from palm oil mill process to achieve river water quality for reuse as recycled water in a zero emission system. J Clean Prod 67:58–61. https://doi.org/10.1016/j.jclepro.2013.12.004
46. Parthasarathy S, Gomes RL, Manickam S (2016) Process intensification of anaerobically digested palm oil mill effluent (AAD-POME) treatment using combined chitosan coagulation, hydrogen peroxide (H_2O_2) and Fenton's oxidation. Clean Technol Environ Policy 18(1):219–230. https://doi.org/10.1007/s10098-015-1009-7
47. Teh CY, Wu TY, Juan JC (2014) Potential use of rice starch in coagulation–flocculation process of agro-industrial wastewater: treatment performance and flocs characterization. Ecol Eng 71:509–519. https://doi.org/10.1016/j.ecoleng.2014.07.005
48. Zahrim AY, Tizaoui C, Hilal N (2010) Evaluation of several commercial synthetic polymers as flocculant aids for removal of highly concentrated C.I. Acid Black 210 dye. J Hazard Mater 182(1–3):624–630. https://doi.org/10.1016/j.jhazmat.2010.06.077
49. Chen X, Wang Z, Fu Y, Li Z, Qin M (2014) Specific lignin precipitation for oligosaccharides recovery from hot water wood extract. Biores Technol 152:31–37. https://doi.org/10.1016/j.biortech.2013.10.113

50. Zahrim AY, Tizaoui C, Hilal N (2011) Coagulation with polymers for nanofiltration pre-treatment of highly concentrated dyes: a review. Desalination 266(1–3):1–16
51. Zinatizadeh AA, Ibrahim S, Aghamohammadi N, Mohamed AR, Zangeneh H, Mohammadi P (2017) Polyacrylamide-induced coagulation process removing suspended solids from palm oil mill effluent. Sep Sci Technol (Philadelphia) 52(3):520–527. https://doi.org/10.1080/01496395.2016.1260589
52. Luo H, X-A Ning, Liang X, Feng Y, Liu J (2013) Effects of sawdust-CPAM on textile dyeing sludge dewaterability and filter cake properties. Biores Technol 139:330–336. https://doi.org/10.1016/j.biortech.2013.04.035
53. Abu-Hasan MA, Puteh MH (2007) Pre-treatment of palm oil mill effluent (POME): a comparison study using chitosan and alum. Malays J Civil Eng 19(2):128–141
54. Shak KPY, Wu TY (2014) Coagulation-flocculation treatment of high-strength agro-industrial wastewater using natural *Cassia obtusifolia* seed gum: treatment efficiencies and flocs characterization. Chem Eng J 256:293–305. https://doi.org/10.1016/j.cej.2014.06.093
55. Birima AH, Ahmed AT, Noor MJMM, Sidek LM, Muda ZC, Wong LS (2015) Application of salt extracted peanut seeds in the pretreatment of palm oil mill effluent (POME). Des Water Treat 55(8):2196–2200. https://doi.org/10.1080/19443994.2014.930696
56. Shak KPY, Wu TY (2017) Synthesis and characterization of a plant-based seed gum via etherification for effective treatment of high-strength agro-industrial wastewater. Chem Eng J 307:928–938. https://doi.org/10.1016/j.cej.2016.08.045
57. Khayet M, Zahrim AY, Hilal N (2011) Modelling and optimization of coagulation of highly concentrated industrial grade leather dye by response surface methodology. Chem Eng J 167:77–83
58. Zahrim AY, Nasimah A, Hilal N (2015) Coagulation/flocculation of lignin aqueous solution in single stage mixing tank system: modeling and optimization by response surface methodology. J Environ Chem Eng 3(3):2145–2154. https://doi.org/10.1016/j.jece.2015.07.023
59. Tamrin KF, Zahrim AY (2017) Determination of optimum polymeric coagulant in palm oil mill effluent coagulation using multiple-objective optimisation on the basis of ratio analysis (MOORA). Environ Sci Pollut Res 24(19):15863–15869. https://doi.org/10.1007/s11356-016-8235-3
60. Mohd Tadza MY, Ghani NAF, Mohammad Sobani HH (2016) Evaluation of sludge from coagulation of palm oil mill effluent with chitosan based coagulant. Jurnal Teknologi 78(5–4):19–22. https://doi.org/10.11113/jt.v78.8529
61. Kuboi T, Fujii K (1985) Toxicity of cationic polymer flocculants to higher plants n hydroponic cultures. Soil Sci Plant Nutr 31(2):163–173. https://doi.org/10.1080/00380768.1985.10557424
62. Kuboi T, Fujii K (1984) Toxicity of cationic polymer flocculants to higher plants: I. Seedling assay. Soil Sci Plant Nutr 30(3):311–320. https://doi.org/10.1080/00380768.1984.10434697
63. Kumar A, Saha A (2011) Effect of polyacrylamide and gypsum on surface runoff, sediment yield and nutrient losses from steep slopes. Agric Water Manag 98(6):999–1004. https://doi.org/10.1016/j.agwat.2011.01.007
64. Lee KE, Morad N, Teng TT, Poh BT (2012) Development, characterization and the application of hybrid materials in coagulation/flocculation of wastewater: a review. Chem Eng J 203:370–386. https://doi.org/10.1016/j.cej.2012.06.109

Effect of Seaweed Physical Condition for Biogas Production in an Anaerobic Digester

N. Bolong, H. A. Asri, N. M. Ismail and I. Saad

Abstract The increasing demand for environmental protection and renewable energy has made bioenergy technologies such as anaerobic digestion substantially attractive. The main objective of this study is to determine the biogas yield from the raw seaweed *Eucheuma cottonii* and waste products using anaerobic digestion, operated under different physical conditions. Seaweeds comprise of a thallus (leaf like) and sometimes a stem and a foot (holdfast). Seaweed has the potential to be developed into the raw and waste material for biogas due to higher growth rates, greater production yields, and higher carbon fixation rates than land crops. Seaweed has 4–39% carbohydrate content and a high moisture content with low lignin compared to other terrestrial plants, thus it is simpler to be degraded. The integration of the findings may be the key to make seaweed waste product that is more efficient and affordable to serve as a sustainable and renewable energy source. The study used 1.5 L anaerobic digesters for fresh and 3-month-old *Eucheuma* sp. evaluated at different stages by monitoring the pH, chemical oxygen demand, and biogas production. The study found that within 18 days, the anaerobic digestion of *E. cottonii* seaweed yielded 0.4–1 ml biogas/g seaweed with up to 56% methane content.

Keywords Anaerobic digestion · Biogas · *Eucheuma cottonii* seaweed

1 Introduction

Biogas is a renewable source of energy that can be harvested to reduce the impact on environment and health of the rural and urban population. The idea of producing gas from extracted waste material could be seen as early as the seventeenth century. It was first discovered by Jon Baptita Van Helmont, who said that organic waste

N. Bolong (✉) · H. A. Asri · N. M. Ismail · I. Saad
Faculty of Engineering, Universiti Malaysia Sabah (UMS),
88400, Kota Kinabalu, Sabah, Malaysia
e-mail: nurmin@ums.edu.my

N. Horan et al. (eds.), *Anaerobic Digestion Processes*,
Green Energy and Technology, https://doi.org/10.1007/978-981-10-8129-3_10

that is in the process of decaying could generate flammable gas; then Count Alessandro Volta continued the study by concluding that the amount of decaying organic matter is directly proportional to the amount of gas produced [1].

The biogas produced usually contains 50–65% methane, 35–50% carbon dioxide [2]. However, the proportions of methane and carbon dioxide vary with the duration and extent of biomethanation over the retention time [3]. Even though all the organic materials degrade and produces biogas in anaerobic digesters, it is still an interesting subject of study due to the complexity of the bioconversion process. The main three parameters that affect anaerobic digester performance are the: (i) feedstock characteristics, (ii) reactor design, and (iii) operational conditions of which temperature and pH are the most important parameters [4].

Even though methane has a commercial value, its emission to the environment causes the greenhouse effect and when compared to natural gas, it has half the calorific values. One of the primary causes of the greenhouse gas emissions of CO_2 and CH_4 release is due to landfilling [5]. Hence, appropriate collection and tapping of the biogas produced are vital to reduce its impact on the environment. Besides, biogas technology also transforms organic waste to high-quality fertilizers [6].

The extensive coastline in Malaysia is surrounded by numerous islands and thus provides habitats for seaweed proliferation. Sabah is one state in Malaysia which is commercially producing seaweeds and this is increasing and highlighted as one of the most important aquaculture commodities. Seaweed has the potential to be developed into the raw and waste material for biogas production. It is a multicellular plant with no roots, with stems, and leaves that grow in salt or fresh water. Seaweed has 4–39% carbohydrate content and a high moisture content with low lignin compared to other terrestrial plants, thus it is easier to degrade [7, 8]. Seaweed does not require land freshwater for cultivation thus does not compete with the growing of food crops or with the residential land, hence the choice of utilizing seaweed algae as a biomass material for biogas is high. Furthermore, new harvesting techniques and valuable co-products produced by some algal strains have been discovered. These improvements have led to a rise in interest for using these organisms for bioenergy generation [9]. The biomethane potential of seaweed is greatly dependent on its chemical composition, which is highly variable due to its type, habitat, cultivation method and time of harvest [10]. It also has a low C/N ratio that might cause problems in the anaerobic digester [11]. Seaweeds have methane yields ranging from 0.14 to 0.40 m^3/kg volatile solids (VS), which is similar to methane production from primary sewage sludge and therefore suitable as a raw material for anaerobic fermentation [12]. Biogas production from seaweed process has been demonstrated to be technically viable, however, the cost of this process is still high and there is a need to reduce the cost of the raw material by at least 75% over current levels before it is competitive in the current market [13]. A recent study by Seghetta et al. [14] reveals that whether it is used for energy via anaerobic digestion or protein production, seaweed provides environmental benefits in terms of mitigation of climate change, with biogas production from dried *Laminaria digitata* being the most favorable scenario.

The edible seaweed *Eucheuma cottonii* is also known as *Kappaphycus alvarezii*. Seaweeds are usually classified into three broad groups which are brown seaweed (*Phaeophyceae*), red seaweed (*Rhodophyceae*), and green seaweed (*Chlorophyceae*) [15]. With over 6000 species, red algae top the algae phylum, followed by brown algae with 2000 species, and green algae with 1200 species [16]. Sabah Malaysia is the main center of seaweed production, with Semporna the most active district [17]. Therefore, the possibility to utilize rich seaweed biomass in Sabah for biogas production is high. This could help curb the main pollution issues while also helping to reduce the usage of commercial fossil fuel by exploring the biogas production from seaweed. But there are challenges in using seaweed for biogas production, such as high water content, variations in nutrient content because of season changes, low C/N ratio [18] and the algal cell walls that can be hard to break down [19]. To tackle these issues, wet anaerobic digestion or other pretreatments were introduced into parts of the systems with modification of carbon-rich materials such as biosludge to reduce the toxicity of ammonia [20].

Consequently, the main objective of this study was to determine and assess the biogas production possibility of *E. cottonii* seaweed using two types of fresh and waste and comparing different physical sizes range of seaweed in an anaerobic digester. Seaweed was chosen due to the huge local availability and feasibility of biogas production feasibility with low carbohydrate and high moisture content, which helps in degrading the biomass into biogas.

2 Materials and Methods

2.1 Substrates and Inoculums

Seaweed biomass or *Eucheuma* sp. were bought from Kota Kinabalu market in Sabah Malaysia. Two types substrates material shown in Fig. 1a and b, which are fresh (green colored) and waste (white pale) seaweeds were stored at room temperature of 23 °C for 3 days and 3 months, respectively, before feeding to the anaerobic digesters.

Horse manure was used as inoculum to digest the energy crops and biomass. Horse manure has a higher carbon and nitrogen content than carbon and nitrogen content in cow manure, which is a source of energy for microorganisms [21]. The feeding took place to catalyze the process of extracting biogas. The manure was locally available and collected from Sabandar Leisure Rides in Tuaran, Sabah, and each 1 kg manure is dissolved in 6 L distilled water to maintain consistency in the digesters.

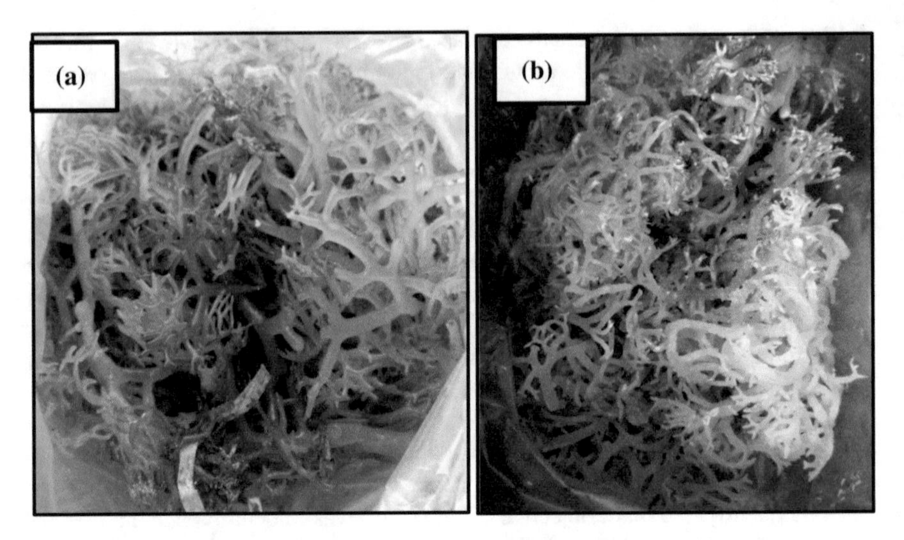

Fig. 1 **a** Raw *Eucheuma cottonii* (3 days); **b** Waste *Eucheuma cottonii* (3 months)

2.2 Reactors and Operations

Six identical reactors with a liquid volume of 1.5 L, labeled as R1, R2, R3 for raw type and W1, W2, and W3 for waste type were equipped with a magnetic stirrer to provide sufficient mixing for substrates. Table 1 summarizes the substrate ID and its main physical characteristic.

The respective digesters were filled with 500 g seaweed substrate and 1 L horse inoculum. The rotation speed was set at a rate of 70 RPM and ran continuously for 18 days. Figure 2 shows the schematic drawing of the anaerobic digester setup used for the experiment. The prototype used was 1.5 L in volume, which was made from the plastic airtight container, connected with a flow meter and balloon for gas collection. The reactor was wrapped with black-painted aluminum foil to trap and maintain heat inside the digester at 29–31 °C (mesophilic condition) using a hot plate magnetic stirrer continuously monitored using a thermometer. The collected biogas was determined by using gas analyzer (GasAlertMax XT II) for the methane

Table 1 Reactor ID and substrates condition

Reactors ID	Substrates type and its physical condition
R1	Raw *Eucheuma cottonii*—Original size
R2	Raw *Eucheuma cottonii*—Cut using scissors into 2–5 cm
R3	Raw *Eucheuma cottonii*—Shredded using blender into 0.5–1 cm
W1	Waste *Eucheuma cottonii*—Original size
W2	Waste *Eucheuma cottonii*—Cut using scissors into 2–5 cm
W3	Waste *Eucheuma cottonii*—Shredded using blender into 0.5–1 cm

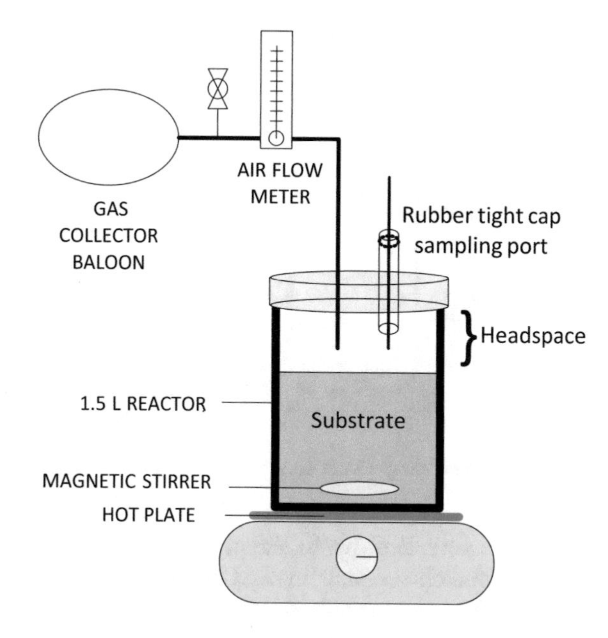

Fig. 2 Schematic of the seaweed anaerobic digester setup

gas, whereas digesters pH was measured by using pH meter (Hanna HO9811-5), and chemical oxygen demand (COD) was determined by using Azide Modification Method (DR890 Colorimeter).

3 Results and Discussions

3.1 Effect of Temperature and on pH Seaweed

The seaweed family of *Solieriaceae*; *E. cottonii*, was studied to generate biogas in an anaerobic digester. The variation of temperature during anaerobic digester was recorded over the range 29–31 °C, ensuring a suitable temperature of mesophilic condition as recorded in Fig. 3. The highest temperature value recorded was on the third (3rd) day of the experiment at 31.2 °C, while the lowest reading was 29.6 °C, obtained on the sixth (6th) day of the experiment. Methane production has been documented over a various range of temperature, but the most productive is either mesophilic conditions, at 30–35 °C or the thermophilic range at 50–55 °C [22]. Furthermore, limitations in a thermophilic digester caused a much longer startup period than a mesophilic digester to allow mesophilic sludge to acclimatize with the substrate as well as the temperature shift [23]. Hence, the seaweed anaerobic digesters were under mesophilic temperature condition as expected. The monitored temperature in this study falls in the mesophilic range of mesophilic with uniform temperature changes less than 2 °C. This is acceptable because the anaerobic

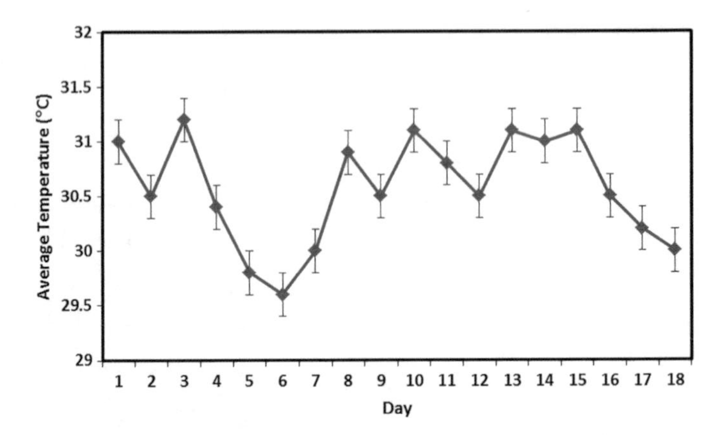

Fig. 3 Variation temperature during anaerobic digester process

process is very sensitive to abrupt temperature changes that may cause unbalance between microbial population and thus the normal limit is about 2 °C per day [24].

The influence of the physical condition of the seaweed substrate in relation to its pH and COD was shown in the Figs. 4 and 5. It was observed that over the initial 4–5 days, the pH was alkaline for all substrates and then dropped to become acidic. Based on the recorded pH, raw seaweed fermented after 5th day whereas waste seaweed earlier on its 2nd day. The pH value varied because of the volatile fatty acids concentration, alkalinity, and buffering capacity of the system [25]. It was observed that the pH would drop gradually to a value of around 6.0. This was due to the early phases of anaerobic digestion, acidogenesis, and acetogenesis. The phases increased the amount of H^+ ions in the digester and prevailed over the methanogenesis phase thus giving a reduction of pH value.

During the second phase of acidogenesis (fermentation) stage, the hydrolysed products are altered to volatile unsaturated fats, alcohols, aldehydes, ketones,

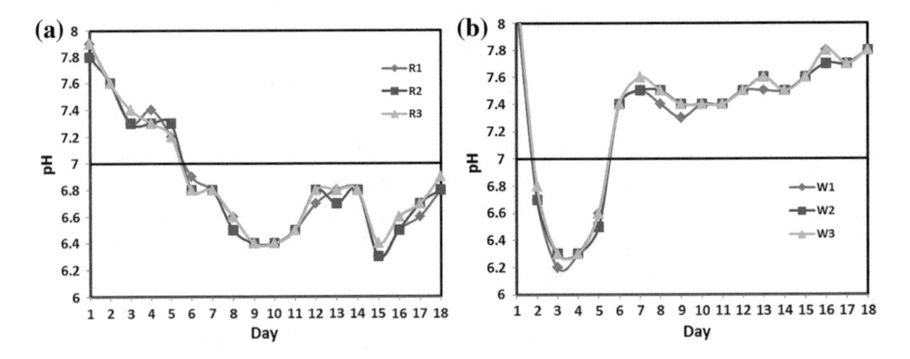

Fig. 4 pH evolution between different stages for **a** raw seaweed and **b** waste seaweed over 18 days in the anaerobic reactor

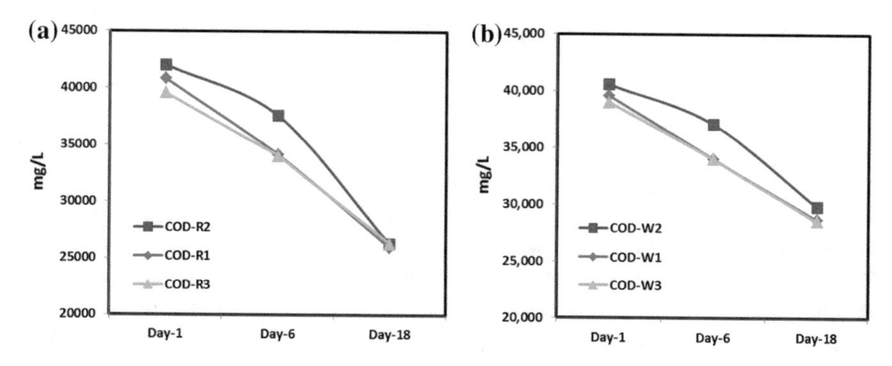

Fig. 5 Chemical oxygen demand (COD) measured against time (day) between **a** raw seaweed and **b** waste seaweed

ammonia, carbon dioxide, water, and hydrogen by the acid-forming bacteria. The organic acids formed are acetic acid, propionic acid, butyric acid, and valeric acid. Volatile fatty acids with more than four-carbon chain cannot be utilized directly by methanogens [26] as observed in the case of raw seaweed type.

The accompanying stage is acetogenesis, where organic acids are further oxidized to acetic acid, hydrogen, and carbon dioxide which are utilized as a part of the following procedure. Acetogenesis likewise incorporates acetic acid generation from hydrogen and carbon dioxide by acetogens and homoacetogens. The transition of the substrate causes the pH of the system to drop which is beneficial to acidogenic and acetogenic organisms as confirmed by Ostrem [27].

The final processes of digestion are methanogenesis. The optimum pH for the methanogenesis stage is a pH between 7.2 and 8.2, if the pH falls below 6, the anaerobic degradation rate will decrease, and this depends on the balanced activity of microorganisms [10]. However, the process can tolerate a pH range of 6.5 up to 8.0 [4]. In this study, the pH after 18 days in the digester showed that waste seaweed (W1, W2, and W3) has undergone methanogenesis beginning on the 6th day and stabilizing near pH 7.7 for the rest of study. Comparatively, raw seaweed (R1, R2, and R3) were still below pH 7 and might continue into the methanogenesis stage after the 15th day. Based on the recorded pH, similar trends were found regardless of the physical condition of seaweed, either in its original size or cut and shredded.

The COD-recorded initially was high in the range of 38,000–42,000 mg/L which may be attributed to the inoculum of horse dung. But the measured COD between raw and waste seaweed reactors was almost identical and gave a consistent outcome. Nevertheless, it was observed that the COD decreased over the 18-day process with an average of 30–40%. For both raw and waste seaweed, the 5 cm size gave a higher COD value and this could be due to the optimum size that consumes oxygen during decomposition as observed from other work [28]. The results were shown in Fig. 5.

3.2 *Effect of Biogas Production Between Raw and Waste Seaweed*

The biogas production is shown in Fig. 6. The volumes are quoted at normal pressure and the temperature found inside the anaerobic tanks. The biogas generation was relatively low at the start, then increased and gradually leveled off for all the samples. In the anaerobic setup, the biogas production started to increase significantly indicating that a high proportion of waste was being broken down into simpler molecules. Biogas production of seaweed within the 18-day testing period was recorded between 200 and 500 mL per 500 g of seaweed samples which is equivalent to 0.4–1 mL/g of seaweed. In spite of variations of species and geographical influences, different pretreatment and longer duration (more than 2 weeks of digestion process) the biogas produced was less than the result of Kawaroe et al. [29] that produced up to 2 mL/g.

Among the three sizes of seaweeds, the smallest of 1 cm shows the highest biogas volume produced. This is because the smaller size of seaweed has a higher surface area that exposes seaweed and allows a faster reaction. In this study, the biogas volume produced was greatly influenced by the physical condition where 1 cm seaweed > 5 cm > original size. Furthermore, the raw type shows a higher biogas volume and it is expected to produce more biogas after 18th day; however, a duration of more than 18 days was not further explored in this work as the biogas produced was targeted to be purified using a membrane process in our future study.

The analysis of biogas composition is illustrated in Fig. 7 and shows the range of 55–57% methane gas for waste-type seaweed. Raw-type biogas products were not measured and reported here due to the longer reaction time needed for the methanogenesis stage as explained previously, based on pH evolution measurements. It should be noted that the biogas produced will also contain mixtures of other gases such as carbon dioxide, hydrogen sulfide, and carbon monoxide [4] which could not be measured due to the restrictions of the gas analyzer used in this work. However, the study found that the physical condition of seaweed whether

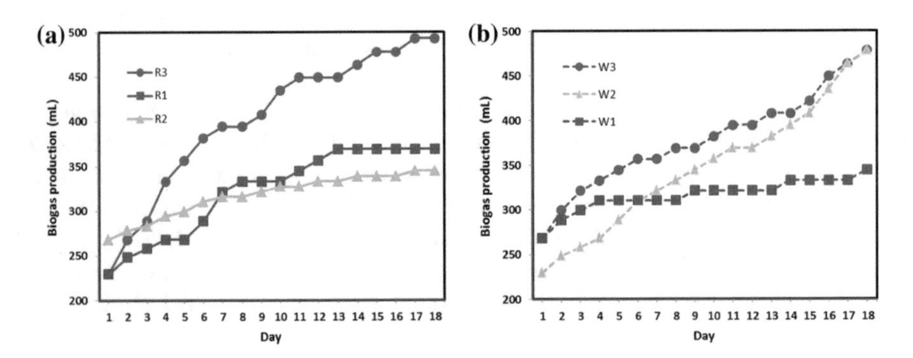

Fig. 6 Biogas volume as a function of time and sizes in **a** raw and **b** waste seaweed type

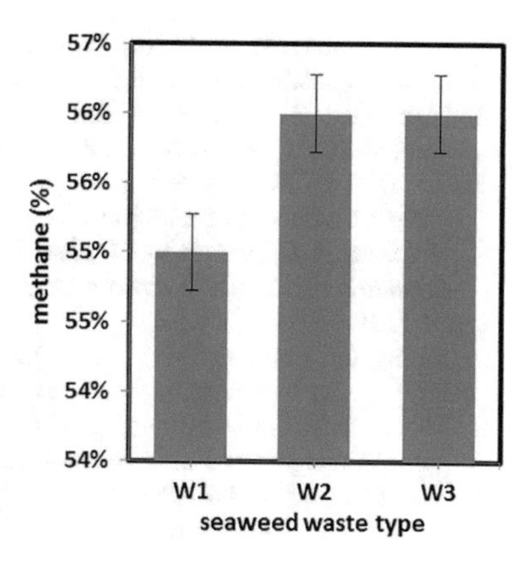

Fig. 7 Methane percentage composition produced by waste seaweed

shredded or the original size did not produce significant differences in methane production. The 500 g waste seaweed which had been kept for 3 months before undergoing anaerobic digestion had produced up to 56% of methane as shown in Fig. 7. The results of biogas produced by seaweed show a positive outcome and consistent with the common composition of 55–60% methane content reported by other works [4, 30] and could become a useful supplement for efficient anaerobic digestion operation [31].

4 Conclusion

A study into the effect of seaweed size of raw and waste biomass of *E. cottonii* was conducted for biogas production. In light of the results obtained, it was concluded that raw and waste (3 months) seaweed produced biogas and the duration or rate of the digestion process was affected by the pH. The seaweed size has only a minor influence in biogas production and chemical oxygen demand during decomposition. This attempt for *E. cottonii* yielded 0.4–1 ml/g biogas with up to 56% methane within 18 days in a 1.5 L anaerobic digester laboratory setup.

References

1. Abbasi T, Tauseef SM, Abbasi SA (2012) A brief history of anaerobic digestion and biogas. Biogas Energy: 11–23
2. World Energy Council (1994) New renewable energy resources: a guide to the future. London Kogan page limited
3. Kavuma C (2013) Variation of methane and carbon dioxide yield in a biogas plant MSc. Thesis report, Royal Institute of Technology, Stockholm, Sweden
4. Ciobla AE, Lonel L, Dumitrel G-A, Popescu F (2012) Comparative study on factors affecting anaerobic digestion of agricultural vegetal residues. Biotechnol Biofuels 5:39

5. Lungkhimba HM, Karki AB, Shrestha JN (2010) Biogas production from anaerobic digestion of biodegradable household wastes. Sci Technol 11(1):167–172
6. Mshandete A, Kivaisi A, Rubindamayugi M, Mattiasson B (2004) Anaerobic batch co-digestion of sisal pulp and fish wastes. Bayero J Bioresour Technol 95:19–24
7. Sitompul JP, Bayu A, Soerawidjaja TH, Lee HW (2012) Studies of biogas production from green seaweeds. Int J Environ Bioenergy 3(3):132–144
8. Vivekanand V, Eijsink VGH, Horn SJ (2012) Biogas production from the brown seaweed *Saccharina latissima*: thermal pretreatment and codigestion with wheat straw. J Appl Phycol 24:1295–1301
9. Mussgnug JH, Klassen V, Schluter A, Kruse O (2010) Microalgae as substrates for fermentative biogas production in a combined biorefinery concept. J Biotechnol 150(1):51–56
10. Tabassum MR, Xia A, Murphy JD (2017) Potential of seaweed as a feedstock for renewable gaseous fuel production in Ireland. Renew Sustain Energy Rev 68(Part 1):136–146
11. Yen HW, Brune DE (2007) Anaerobic co-digestion of algal sludge and waste paper to produce methane. Bioresour Technol 98(1):130–134
12. Reith JH, Huijgen W, Van Hal J, Lenstra J (2009) Seaweed potential in the Netherlands. Macroalgae - Bioenergy Research Forum, Plymouth, UK
13. Bruton T, Lyons H, Lerat Y, Stanley M, Rasmussen MB (2009) A review of the potential of marine algae as a source of biofuel in Ireland. Sustainable Energy Ireland, Dublin, Ireland
14. Seghetta M, Romeo D, D'Este M, Alvarado-Morales M, Angelidaki I, Bastianoni S, Thomsen M (2017) Seaweed as innovative feedstock for energy and feed—evaluating the impacts through a Life Cycle Assessment. J Clean Prod 150:1–15
15. Mohamed S, Hashim SN, Abdul Rahman H (2012) Seaweeds: a sustainable functional food for complementary and alternative therapy. Trends Food Sci Technol 2(2):83–96
16. Venugopal V (2011) Polysaccharides from seaweed and microalgae marine polysaccharides food applications. CRC Press, Boca Raton, USA, pp 89–129
17. Sade A, Ali I, Ariff MRM (2006) The seaweed industry in Sabah, East Malaysia. Jati-J SE Asian Stud 11:97–107
18. Montingelli ME, Tedesco S, Olabi AG (2015) Biogas production from algal biomass: a review. Renew Sustain Energy Rev 43:961–972
19. Ramos-Suárez JL, Carreras N (2014) Use of microalgae residues for biogas production. Chem Eng J 242:86–95
20. Wang M, Sahu AK, Rusten B, Park C (2013) Anaerobic co-digestion of microalgae *Chlorella* sp. and waste activated sludge. Bioresour Technol 142:585–590
21. Kusch S, Oechsner H, Jungbluth T (2008) Biogas production with horse dung in solid-phase digestion systems. Bioresour Technol 99(2008):1280–1292
22. Choorit W, Wisarnwan P (2007) Effect of temperature on the anaerobic digestion of palm oil mill effluent. Electron J Biotechnol 10(No. 3, Issue of July 15):376
23. Poh EP, Chong MF (2010) Thermophilic palm oil mill effluent (POME) treatment using a mixed culture cultivated from POME. Chem Eng Trans 21:811–816
24. Marcos VS (2005) Biological wastewater treatment in warm climate regions. IWA Publishing, London
25. Abdulkarim BI (2017) Anaerobic digestion: an increasingly acceptable treatment option for organic fraction of municipal solid waste. Int J Sci Res Publ 7(1):79–82
26. Wang Y, Kuninobu M, Ogawa HI, Kato Y (1999) Degradation of volatile fatty acids in highly efficient anaerobic digestion. Biomass Bioenergy 16:407–416
27. Ostrem K (2004) Greening waste: anaerobic digestion for treating the organic fraction of municipal solid wastes. Master thesis, Columbia University
28. Hajji A, Rhachi M (2013) The influence of particle size on the performance of anaerobic digestion of municipal solid waste. Energy Proc 36(2013):515–520
29. Kawaroe M, Slundik, Wahyudi R, Lestari DF (2017) Comparison of biogas production from macroalgae *Eucheuma cottonii* in anaerobic degradation under different salinity conditions. World Appl Sci J 35(3):344–351

30. Deublein D, Steinhauser A (2008) Biogas from waste and renewable resources. Wiley-VCH Verlag GmbH & Co. KGaA, Weinheim
31. Kuroda K, Akiyama Y, Keno Y, Nakatani N, Otsuka K (2014) Anaerobic digestion of marine biomass for practical operation. J Marine Sci Technol 19:280–291

Phosphorus Recovery from Anaerobically Digested Liquor of Screenings

N. Wid

Abstract Phosphorus is a limited resource which is predicted to get exhausted at some point during the twenty-first century. However, it is present in wastewaters at concentrations that come close to supplying the nation's annual requirements for fertiliser. Many papers have addressed the recovery of phosphorus as struvite (magnesium ammonium phosphate hexahydrate) from different types of waste while the most prominent usage of struvite is as a slow-release fertiliser, suitable as a replacement for chemical fertiliser, for agricultural application. In this study, screenings produced during the wastewater treatment process were anaerobically digested to obtain anaerobically digested liquor which was subsequently used for phosphorus recovery in the form of struvite. This was carried out at different concentrations of dry solids. The amount of struvite potential was calculated theoretically using molar ratio calculations of 1:1:1 (Mg:N:P). From the results, it was found that the digestate is high in phosphorus content and can be recovered up to 41%. For struvite yield, 0.27 kg of struvite can be recovered from each kg dry solids of screenings from 3% of dry solids. Screenings thus prove a valuable source of additional phosphorus which current disposal practices fail to exploit.

Keywords Anaerobic digestion · Anaerobically digested liquor of screenings
Phosphorus recovery

1 Introduction

Phosphorus is a limited resource which is anticipated to get exhausted in the twenty-first century. But ironically, in the living environmental system, much phosphorus is discharged and as a result, eutrophication has become a serious problem in receiving waters with a resultant deterioration of water quality. Screenings are produced during the first stage of sewage treatment process in a

---

N. Wid (✉)
Faculty of Science and Natural Resources, Universiti Malaysia Sabah, Jalan UMS, 88400
Kota Kinabalu, Sabah, Malaysia
e-mail: newati@ums.edu.my

N. Horan et al. (eds.), *Anaerobic Digestion Processes*,
Green Energy and Technology, https://doi.org/10.1007/978-981-10-8129-3_11

Table 1 Selected studies on recovery of phosphorus in the form of struvite from different type of wastes

Type of waste	References
Swine wastewater	[3–9]
Municipal wastes	[10–19]
Urine	[20, 21]
Synthetic wastewater	[22–25]

wastewater treatment plant and refer to material that may cause operational failure if it passes through mechanical equipment. Screenings comprise rags, paper, plastic, grit, grease, sand and wood, so it is removed on inlet screens that typically have apertures of 6 mm. In the UK alone 150,000 dry tonnes of screenings are produced every year with the majority of this disposed to landfill [1]. As it contains a substantial amount of nutrient with high organic content (>90%), the disposal options may lead to environmental issues such as limited space remaining for landfilling, odour problems, greenhouse gases (GHGs) emissions and nutrient release. Therefore, application of anaerobic digestion is important as it not only alleviates environmental problems, but also helps to reduce the stress on depleted resources [2]. Anaerobic digestion of screenings produces a liquid or biosolid mixture which is high in nitrogen and phosphorus termed here as anaerobically digested liquor. Anaerobic digestion will solubilise the nutrient and make them easy to recover as valuable fertiliser components. The nutrient-rich effluent can be passed forward for precipitation to recover phosphorus in a reusable form of fertiliser. Precipitation of phosphorus is a controlled process that is highly dependent on the physical–chemical parameters of the solution, prominently pH. Most of the phosphate salts produced from the precipitation technique is in the form of struvite, a white substance that shows potential as a slow-release fertiliser. The potential of struvite precipitation can also be calculated theoretically using molar ratio. By recovering phosphorus from anaerobically digested liquor of screenings, this can turn waste into useful resources. A number of studies have been reported in using anaerobic-digested liquor from various types of wastes to recover phosphorus (Table 1). However, a study on anaerobic digestion of screenings is very rare as this material has received little attention. Therefore, this study aims to investigate the potential of phosphorus recovery from screenings by performing anaerobic digestion to obtain the anaerobically digested liquor. The struvite potential was then determined using the molar ratio of magnesium, ammonium and phosphate present in the digested liquor.

2 Materials and Methods

2.1 Anaerobically Digested Liquor

The anaerobically digested liquor in this study was obtained from batch anaerobic reactor digesting screenings as feedstock for 30 days, as described in

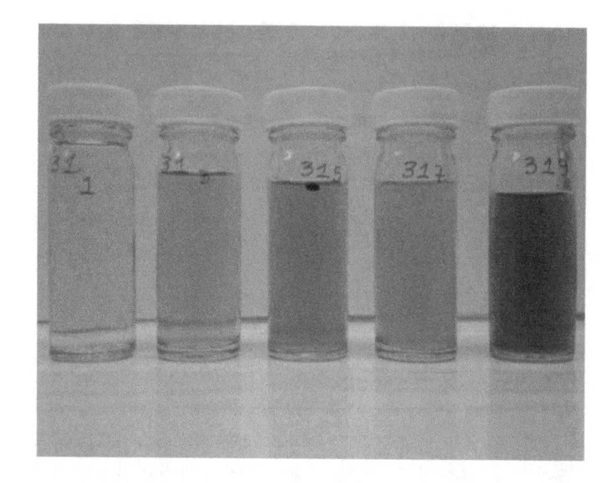

Fig. 1 Anaerobically digested liquor of screenings

Chap. "Microbial Fuel Cell (MFC) Development from Anaerobic Digestion System " (Sect. 2.4). To study the potential of phosphorus recovery, the digested effluent produced on the final day was centrifuged at 4000 rpm for 30 min and filtered through a 0.45 μm membrane filter (Fig. 1). The resultant supernatant was analysed for its magnesium, ammonium and phosphate concentrations according to the procedures outlined in [26]. This procedure was repeated for all digested liquor produced from each reactor, labelled as R1, R2, R3 and R4, representing different concentrations of dry solids.

2.2 *Struvite Potential in Phosphorus Recovery*

The potential of struvite precipitation was calculated theoretically according to the method described by [16]. Struvite potential from the digested liquor of different concentrations of dry solids was calculated using a molar ratio of magnesium, ammonium and phosphate. Struvite has the chemical formula $MgNH_4PO_4 \cdot 6H_2O$ and reacts in 1:1:1 molar ratio to form struvite. The molecular weight of struvite is 244 g/mol. Theoretical weight of struvite that can be calculated using Eq. 1.

$$S = \text{number of moles} \times \text{MWS} \tag{1}$$

where

S weight of struvite (g/L),
MWS molecular weight of struvite (244 g/mol).

The number of moles is the minimum number of available moles in the effluent (among Mg^{2+}, NH_4^+ and PO_4^{3-}) which was used in the formation of struvite.

Table 2 Phosphorus recovery at different concentrations of dry solids

Reactor (% dry solids)	P recovery (%)	Struvite yield (kg/kg dry solids)
R1 (3)	37	0.27
R2 (6)	41	0.19
R3 (9)	39	0.18
R4 (12)	38	0.15

3 Results and Discussion

The potential of phosphorus recovery was evaluated based on the percentage of phosphorus produced in the digestate from dry solids of screenings during the digestion. The phosphorus recovery was also expressed in struvite yield, which was calculated theoretically. The results show not much variation in the percentage when studied at different dry solids concentrations. It ranged between 37 and 41%, with 6% of dry solids shows the highest. The results indicate that screenings provide an ideal feedstock for phosphorus recovery. When phosphorus was recovered in the form of struvite, using unit kg of struvite/kg dry solids of screenings, 3% of dry solids produced the highest phosphorus with 0.27 kg/kg dry solids, followed by 6, 9 and 12% (Table 2). It suggests the struvite yield decreasing with an increase of the dry solids concentrations. This may due to overloading of organic acids in the digester that may upset the phosphorus release. Considering the amount of screenings produced in the UK yearly, approximately 40,500 tonne of struvite can be recovered per year, by diverting screenings from the landfill to perform anaerobic digestion.

4 Conclusions

This study was developed to investigate the potential of phosphorus recovery from screenings, a difficult heterogeneous type of waste. Anaerobic digestion was performed at controlled pH and temperature for 30 days to solubilise magnesium, ammonium and phosphate ions that involve the precipitation of struvite. Anaerobically digested liquor produced after digestion was rich in nutrient content suggests that screenings is an ideal feedstock for phosphorus recovery. In this study, the struvite potential was calculated theoretically by knowing the concentrations of ions involved in a unit molar, using 1:1:1 molar ratio of Mg:N:P. The results indicate that sufficient phosphorus was released in the digested liquor with the highest was 41% P recovery from 6% dry solids, with small variations on the percentages. This study also suggests the lowest dry solids, i.e. 3%, produces the highest struvite yield with 0.27 kg struvite/kg dry solids of screenings. However, higher dry solids concentrations do not favour struvite yield. Therefore, when using difficult waste such as screenings in recovering phosphorus, it is suggested to perform anaerobic digestion at lower dry solids to optimise phosphorus recovery in the form of struvite. By performing anaerobic digestion, it not only produces an easy reusable form of phosphorus with excellent fertiliser quality, that is a potential source of revenue, but also reduces sludge generation in wastewater treatment plants as well as partially offsetting the cost of treatment.

References

1. Horan NJ (2008) Design and operation of wastewater treatment plants for freshwater fisheries directive (FFD) compliance. A. E. T. Transfer., Ed., ed. UK
2. Tyagi VK, Lo S-L (2013) Sludge: a waste or renewable source for energy and resources recovery? Renew Sustain Energy Rev 25:708–728
3. Huang H, Liu J, Wang S, Jiang Y, Xiao D, Ding L et al (2016) Nutrients removal from swine wastewater by struvite precipitation recycling technology with the use of $Mg_3(PO_4)_2$ as active component. Ecol Eng 92:111–118
4. Huang H, Zhang P, Zhang Z, Liu J, Xiao J, Gao F (2016) Simultaneous removal of ammonia nitrogen and recovery of phosphate from swine wastewater by struvite electrochemical precipitation and recycling technology. J Cleaner Prod 127:302–310
5. Song Y, Dai Y, Hu Q, Yu X, Qian F (2014) Effects of three kinds of organic acids on phosphorus recovery by magnesium ammonium phosphate (MAP) crystallization from synthetic swine wastewater. Chemosphere 101:41–48
6. Liu Y, Kwag J-H, Kim J-H, Ra C (2011) Recovery of nitrogen and phosphorus by struvite crystallization from swine wastewater. Desalination 277:364–369
7. Nelson NO, Mikkelsen RL, Hesterberg DL (2003) Struvite precipitation in anaerobic swine lagoon liquid: effect of pH and Mg:P ratio and determination of rate constant. Bioresour Technol 89:229–236
8. Huang H, Xu C, Zhang W (2011) Removal of nutrients from piggery wastewater using struvite precipitation and pyrogenation technology. Bioresour Technol 102:2523–2528
9. Jordaan EM, Ackerman J, Cicek N (2010) Phosphorus removal from anaerobically digested swine wastewater through struvite precipitation. Water Sci Technol 61:3228–3234
10. Yoshino M, Yao M, Tsuno H, Somiya I (2003) Removal and recovery of phosphate and ammonium as struvite from supernatant in anaerobic digestion. Water Sci Technol 48:171–178
11. Battistoni P, Fava G, Pavan P, Musacco A, Cecchi F (1997) Phosphate removal in anaerobic liquors by struvite crystallization without addition of chemicals: preliminary results. Water Res 31:2925–2929
12. Booker NA, Priestley AJ, Fraser IH (1999) Struvite formation in wastewater treatment plants: opportunities for nutrient recovery. Environ Technol 20:777–782
13. Münch EV, Barr K (2001) Controlled struvite crystallisation for removing phosphorus from anaerobic digester sidestreams. Water Res 35:151–159
14. Pastor L, Mangin D, Ferrer J, Seco A (2010) Struvite formation from the supernatants of an anaerobic digestion pilot plant. Bioresour Technol 101:118–125
15. Song Y-H, Qiu G-L, Yuan P, Cui X-Y, Peng J-F, Zeng P et al (2011) Nutrients removal and recovery from anaerobically digested swine wastewater by struvite crystallization without chemical additions. J Hazard Mater 190:140–149
16. Celen I, Turker M (2001) Recovery of ammonia as struvite from anaerobic digester effluents. Environ Technol 22:1263–1272
17. Talboys PJ, Heppell J, Roose T, Healey JR, Jones DL, Withers PJA (2016) Struvite: a slow-release fertiliser for sustainable phosphorus management? Renew Sustain Energy Rev 26:708–728

18. Kataki S, West H, Clarke M, Baruah DC (2016) Phosphorus recovery as struvite from farm, municipal and industrial waste: feedstock suitability, methods and pre-treatments. Waste Manage 49:437–454
19. Wid N, Selaman R, Jopony M (2017) Enhancing phosphorus recovery from different wastes by using anaerobic digestion technique. Adv Sci Lett 23:1437–1439
20. Zang G-L, Sheng G-P, Li W-W, Tong Z-H, Zeng RJ, Shi C et al (2012) Nutrient removal and energy production in a urine treatment process using magnesium ammonium phosphate precipitation and a microbial fuel cell technique. Phys Chem Chem Phys 14:1978–1984
21. Wilsenach JA, Schuurbiers CAH, van Loosdrecht MCM (2007) Phosphate and potassium recovery from source separated urine through struvite precipitation. Water Res 41:458–466
22. Song Y, Yuan P, Zheng B, Peng J, Yuan F, Gao Y (2007) Nutrients removal and recovery by crystallization of magnesium ammonium phosphate from synthetic swine wastewater. Chemosphere 69:319–324
23. Le Corre KS, Valsami-Jones E, Hobbs P, Parsons SA (2005) Impact of calcium on struvite crystal size, shape and purity. J Cryst Growth 283:514–522
24. Pastor L, Mangin D, Barat R, Seco A (2008) A pilot-scale study of struvite precipitation in a stirred tank reactor: conditions influencing the process. Bioresour Technol 99:6285–6291
25. Ali MI, Schneider PA (2006) A fed-batch design approach of struvite system in controlled supersaturation. Chem Eng Sci 61:3951–3961
26. APHA (2005) Standard methods for the examination of water and wastewater. American Public Health Association, Washington D.C.

Printed in the United States
By Bookmasters